NEUROMARKETING AI

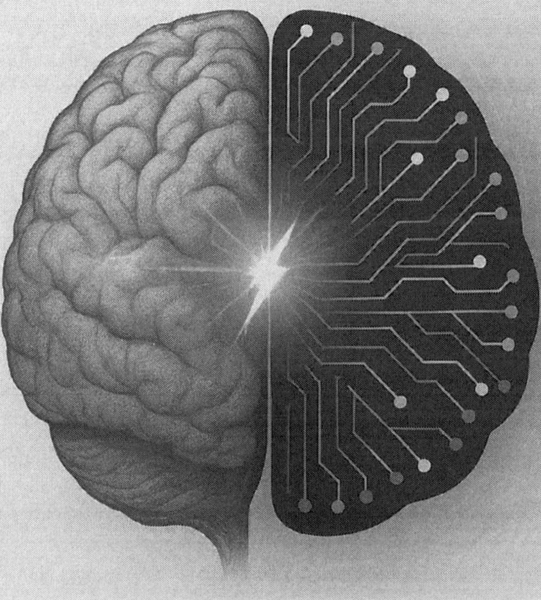

THE NEW SCIENCE OF PERSUASION IN THE AGE OF INTELLIGENT MACHINES

CHRISTOPHE MORIN, PH.D.

CO-AUTHOR OF THE *PERSUASION CODE & NEUROMARKETING*

AUTHOR OF THE *ULTIMATE NEUROMARKETING GUIDE*

Abstract

In an era where attention is the scarcest resource, *Neuromarketing AI* offers a revolutionary framework for decoding and influencing human behavior by aligning ancient neural instincts with modern artificial intelligence. This book synthesizes decades of research in neuroscience, psychology, and marketing, along with the latest advances in AI, to create a persuasive blueprint that is both scientifically grounded and ethically sound.

At its core lies the premise that the primal brain, our evolutionary command center for survival, drives most of our decisions long before logic gets a vote. Traditional marketing strategies, fixated on conscious thought and rational appeals, miss this fundamental truth. *Neuromarketing AI* introduces readers to a neuro-persuasive arsenal: the primal-rational brain model, the dual-system theory of cognition, and the NeuroMap® persuasion framework.

These models demystify why emotion trumps logic and why stimuli must first pass the primal gatekeeper to earn attention and trust. Through case studies and empirical evidence, the book illustrates how marketers can utilize biometric tools, such as EEG, GSR, eye tracking, and facial coding, among others, to measure unconscious responses with precision. It critiques outdated methods, including focus groups and surveys, revealing how neuromarketing tools uncover what people truly feel but cannot articulate.

Finally, the book explores how AI can simulate and scale the mechanisms of persuasion. It outlines how large language models, when properly prompted, can generate content that resonates with the brain's operating system. Readers learn the principles of NeuroPrompting: how to engineer AI inputs that speak to what the primal brain craves. It also presents a roadmap to building custom GPTs that can analyze, predict, and optimize marketing campaigns based on neural principles.

Ethical concerns are woven throughout: who owns brain data, how to avoid manipulation, and how to balance influence with integrity. The book concludes by looking ahead to brain-computer interfaces and the evolving responsibilities of marketers in a hyperconnected, neuro-aware world.

I am incredibly grateful to Dr. Bonnie Bright for her help, guidance, and exceptional editing contributions.

I dedicate this book to the love of my life, my sons, mother, father, brother, sisters, and all the beautiful souls who have supported my splendid life journey.

To Patrick Renvoisé, my co-founder at SalesBrain: thank you for helping transform a radical idea into a life's work. Your partnership has shaped this journey in every way.

SBN: 978-0-9979550-9-5

Praise for Neuromarketing AI

Christophe does not just explain neuromarketing, he makes you feel its power. This book is about trust, meaning, and the human spark behind every choice we make. It is rare to find work that connects science and marketing so seamlessly, and that is precisely why it matters.

Angela Cretu
Ex-Global Chief Executive Officer of AVON

Christophe has a rare gift for making the complexities of the human brain accessible and practical. "Neuromarketing AI" builds on his groundbreaking work in The Persuasion Code and extends it into the age of artificial intelligence. A must-read for anyone serious about persuasion, marketing, or the future of influence.

Ian McCulloh, Ph.D.
Director, AI Executive & Professional Education
Johns Hopkins University

Christophe's ability to combine deep science with accessible frameworks has helped shape not just a discipline, but a whole movement. Few have done more to bring the brain into the boardroom than Christophe, and with this book, he continues that legacy.

Dr. Thomas Zoëga Ramsøy
Best-selling Author of *Introduction to Consumer Neuroscience*
Founder and CEO of Neurons Inc

Christophe makes the science of persuasion both rigorous and usable. "Neuromarketing AI" bridges neuroscience and artificial intelligence in a way that shows not only how influence works, but why responsibility and trust must anchor it. In a time when AI promises scale but often misses human connection, Christophe offers a framework that is both practical and ethical. This is an important read for marketers, researchers, and leaders shaping the future of human behavior.

Naomi Grewal, Ph.D.
Director of UX Research, LinkedIn
Faculty, University of Southern California

Praise for Neuromarketing AI

Christophe powerfully fuses decades of research with real-world practice to tackle today's defining frontier: where neuromarketing meets AI. With scientific rigor and true plug-and-play clarity, he shows how the primitive brain drives attention and choice with proven frameworks—NeuroMap®, the 12 primary triggers, and NeuroPrompting—to turn insight into immediate action. Christophe's craftsmanship shines on every page: precise, practical, and deeply engaging. A must-have for anyone serious about persuasion.

Vincenzo Russo, Ph.D.
Associate Professor of Consumer Psychology and Neuromarketing
Founder and Director of the Center of Research in Neuromarketing
University of Milano (IULM), Italy.

The Ripple Plan™ would not have come to life this quickly without neuromarketing AI. The Custom GPT Christophe developed gave us the ability to communicate with precision — not just what we wanted to say, but what Donors needed to hear. It transformed a complex idea into a simple, emotionally compelling story that inspired immediate action. In today's noisy world, that kind of clarity is not just valuable, it is game-changing.

Rob Fink, J.D.
CEO of RAF Strategies LLC, Creator of The Ripple Plan

This book is a fascinating journey through the intersection of neuroscience, marketing, and artificial intelligence, offering a fresh blueprint for persuasive communication in this rapidly advancing digital era. Blending rich case studies with leading-edge research, Dr. Morin reveals why emotional resonance and primal brain triggers drive consumer decisions, far more powerfully than conscious logic or self-reported preferences.

Kimberly Rose Clark, Ph.D.
Lead Researcher for the DIRECT Lab
Dartmouth College

Foreword by Dr. Thomas Zoëga Ramsøy

When Christophe Morin and I first crossed paths more than a decade ago, neuromarketing was still a daring idea. The notion that we could apply the tools of neuroscience to understand and improve communication struck many as provocative! It was considered too scientific for marketers, too commercial for scientists. Yet it was precisely in that tension that a new field was born.

Over the years, I have had the privilege of witnessing Christophe's relentless pursuit of clarity in this space. His ability to combine deep science with accessible frameworks has helped shape not just a discipline, but a whole movement. Few have done more to bring the brain into the boardroom than Christophe, and with this book, he continues that legacy.

Neuromarketing AI is both timely and urgent. We stand at the crossroads of two revolutions: one in brain science, the other in artificial intelligence. Christophe bridges these worlds with a rare mix of intellectual rigor and ethical sensitivity. He reminds us that persuasion is not about manipulation, but about aligning the message with the mind, between influence and integrity.

In my own work at Neurons, I have seen how far this vision can be taken when science is done right. We have shown that it is not only possible to measure, but to predict neural responses, such as attention, emotion, cognition, and memory, using carefully validated models. Building on this foundation, we have advanced from predictive models to suggestive AI systems and now into generative models, pushing the boundaries of what, not long ago, might have sounded like science fiction...or even magic. This trajectory is proof that the field Christophe helped pioneer is not just alive but accelerating into new frontiers.

This book does more than explain why the deeply nonconscious brain matters in decision-making. It shows us how the convergence of neuroscience and AI can (and must) be used to craft communication that is more empathetic, more effective, and more responsible. That is right up my alley, as I believe neuromarketing, as part of marketing, serves a purpose in providing meaningful and engaging information from companies to prospective customers, thus serving a societal function. Christophe's concept of NeuroPrompting, for example, is not just a technical innovation. It is an ethical call to design AI systems that respect the very organs they aim to persuade.

Reading these pages, I was struck not only by the clarity of the science but by the wisdom of the vision. In an age of noise, Neuromarketing AI offers a compass. It asks us, as practitioners, scholars, and citizens, to decide what kind of influence we want to unleash into the world. This is the call we need!

Christophe and I have long shared the belief that understanding the brain is not an end in itself. It is a responsibility. With this book, Christophe hands you not just knowledge, but also that responsibility. My hope is that you take it seriously, because the future of persuasion and engagement, and perhaps of trust itself, depends on it.

Dr. Thomas Zoëga Ramsøy
September 20, 2025

Table of Contents

INTRODUCTION

Figure 1: Decoding Human Behavior

What drives us to buy, to click, to say yes? The answers lie not in our conscious thoughts but in the shadows of our brains. For over two decades, I have been possessed by a singular obsession: decoding human behavior through the lens of neuroscience. This journey has led me through research labs, corporate boardrooms, and thousands of hours crunching and interpreting brain data from real people responding to real marketing stimuli. The result is a fusion I call "Neuromarketing AI", a framework that marries the oldest instincts of the human brain with the most cutting-edge tools of artificial intelligence.

Neuromarketing was born out of frustration. Traditional marketing relies on what people say they think and feel. But my research and experience in the field confirm what every parent of a toddler already knows: people often cannot explain their decisions, not because they are deceptive, but because most choices happen beneath the level of our awareness. The rational brain may justify, but the primal brain decides.

AI, meanwhile, offers the power to detect and decode patterns far beyond human perception. Nevertheless, AI, for all its promise, lacks a brain. It does not feel pain. It does not desire. It does not persuade. That is why, when used in isolation, AI-driven content often feels synthetic, forgettable, and emotionally flat.

This book bridges that gap. It is a map for marketers, designers, researchers, and executives who want to stop guessing and start influencing in ways that are ethical, evidence-based, and radically more effective. At its core, it is about teaching machines the language of the part of the brain that holds the key to our attention, emotions, and cognitive engagement: It is called the primal brain. In the coming pages, you will:

- Discover how the primal brain drives decision-making and how 12 triggers can hijack its fight-or-flight attention system.
- Learn why emotion always trumps logic in persuasion.
- Explore how to evaluate marketing messages using biometric and neurophysiological tools.
- Understand the ethical tightrope of influencing human behavior with brain-based methods.
- Use GenAI not just to generate content, but to simulate, test, and refine persuasion.
- Build your AI model by customizing a GPT model trained to recognize what your buyer's brain craves.

You will also encounter technological methods such as EEG, GSR, and fMRI, as well as AI tools that track eye movements and decode facial expressions. However, do not worry — this is not a textbook. Think of it as a guided tour through the fascinating corridors of the brain, illuminated by data and powered by empathy.

We will decode the primal-reptilian brain, that ancient sentinel of survival that still wields enormous influence over what we notice, what we fear, and what we buy. We will delve into decision science, exploring how our frustrations, rewards, and tangible gains activate our most fundamental neural circuits. Furthermore, we will decode the paradox of choice, demonstrating how excessive logic can create paralysis rather than persuasion.

Figure 2: Manipulation

Let me also be clear: *Neuromarketing AI* is not about manipulation. It is about alignment. It is about crafting messages that honor the brain's design rather than fighting against it. When you grasp how the brain filters, ignores, or craves stimuli, you are tuning into the brain's frequency with a calibrated neural dial.

We are living through the most consequential shift in communication since the advent of the printing press. Generative AI can now create content on an unprecedented scale, but only when guided by neuroscience can it connect at a truly human level. This constructive collaboration of ancient biology and modern computation is what will define the next era of marketing.

Let me take you back to Parthenay, France, where my fascination with how people make decisions began long before I heard the word "neuroscience." As a young boy, I watched my father, a self-taught inventor, build mobile shops from scratch—vehicles that brought goods to the market squares of small villages. However, it was not the engineering that fascinated me most. It was watching people react. Some walked by, uninterested. Others stopped, brows raised, drawn in by a word, a color, a scent. What made the difference?

That question never left me. It followed me through multiple degrees in marketing, through boardrooms where I served as CEO and CMO, and into the depths of doctoral research on media psychology and neuroscience. The more I studied the brain, the clearer it became: the key to persuasion lies in pre-conscious processes. And that is where neuromarketing finds its power.

When I co-founded SalesBrain in 2002 with Patrick Renvoisé, the term "neuromarketing" was barely in existence. Our mission felt radical: to apply brain science directly to marketing strategy was not a popular idea at the beginning of our business journey. Nonetheless, we developed a persuasion model called NeuroMap®—a framework that has now been taught to over 250,000 executives in more than 30 countries. The promise of this work was clear: measurable, repeatable improvements in message effectiveness, backed by data.

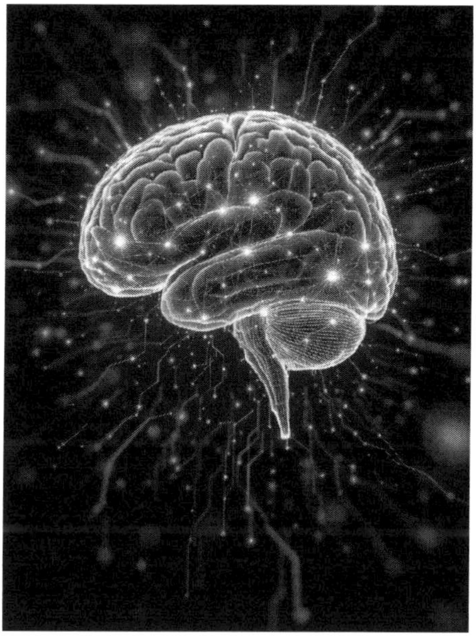

Figure 3: Lighting the Brain

We were venturing into uncharted neural territory, armed with little more than a conviction: that the key to persuasion did not lie in what people *said* they wanted, but in how their brains *light up* when it is engaged. At the time, marketing was still anchored in focus groups and survey logic, tools that asked the rational brain for insights it never owned.

The idea of using biometric sensors, eye tracking, or EEG to decode customer attention and emotional engagement seemed more science fiction than sales strategy. But as we began integrating neuroscience with message testing, a new map emerged—one that replaced guesswork with evidence and illuminated the primal triggers of choice.

Back then, neuromarketing felt like alchemy. Today, it is a science with measurable ROI and ethical imperatives. Furthermore, it all started with one question:

What if the brain had its language of persuasion—and we had not learned to speak it yet?"

But then AI arrived. Initially, it was a tool designed to speed up content creation. Then it became something more: a means to analyze, personalize, and even emulate neural patterns in content generation. More questions started to haunt me.

One in particular:

What if you could build a GPT not just to write, but to influence in ways aligned with how the human brain works?

That is precisely what Neuromarketing AI makes possible!

This book is a synthesis of decades of research and a blueprint for the future. You will explore the neural wiring that governs attention, emotion, and memory. You will discover the biometric tools that reveal what people cannot articulate. You will learn how to engineer prompts that elicit emotionally resonant outputs.

Meanwhile, we will confront the ethical implications of these new capabilities, as decoding the brain also entails protecting it. Persuasion is power. However, it is also a responsibility. In the wrong hands, neuromarketing AI could be used to manipulate, to deceive, and to nudge people toward choices that undermine their autonomy. That is why this book will never shy away from tough questions: Who owns brain data? How is consent defined in the context of predictive modeling? How can we safeguard human dignity for the future?

We are also living in a moment of unprecedented attention scarcity. The average adult receives thousands of branded messages per day, and the human brain has become a master at filtering them out. The result? A crisis of communication where relevance is rare and resonance is rarer still. Neuromarketing AI promises to reclaim attention—not by shouting louder, but by speaking smarter, by delivering signals that the primal brain recognizes as essential: safety, novelty, emotion, and clarity.

This means rethinking everything from campaign planning to UX design. It means measuring not just what your audience clicks, but what their brain remembers and reacts to. Moreover, it means moving past outdated segmentation based on demographics, toward a more dynamic understanding of neural archetypes and emotional states. We will explore how real-time data from facial coding or skin conductance can tell you more about your buyer's intent than any survey or A/B test ever could.

Whether you are creating a campaign, designing a product, writing copy, or managing a brand, set aside your assumptions about persuasion and let us rewire how the world communicates, one synapse at a time.

CHAPTER 1: The History of Persuasion

Persuasion is clearly a sort of demonstration since we are most fully persuaded when we consider a thing to have been demonstrated.

Aristotle

My passion for decoding human behavior has always centered on one driving question: What moves us? Over the course of decades of research, practice, and teaching, I have come to believe that the most potent answer is *persuasion.* Persuasion is more than a marketing tactic; it is the invisible force shaping our relationships, societies, and choices. From whom we trust to what we buy, from how we vote to why we click, persuasion is the thread that weaves through the fabric of decision-making. It is the ultimate interface between mind and message.

To appreciate the power of persuasion in its modern context, it is helpful to trace its intellectual timeline, marked by philosophers, psychologists, economists, and neuroscientists, each contributing to a more nuanced understanding of how influence shapes behavior.

4th Century BCE: Classical Rhetoric

Figure 4: Aristotle

Persuasion, as a discipline, began not in a boardroom or branding meeting but in the agora of ancient Greece. In this bustling civic heart, citizens gathered to trade goods, debate politics, and deliberate on matters of justice. Beyond a simple marketplace, the agora served as the heart of Athenian democracy, where free men expressed their opinions, defended proposals, and influenced public sentiment. It was in this crucible of civic life that the art of rhetoric was born. Among its most eloquent promoters was the philosopher Aristotle, who distilled the chaotic interplay of speeches and debates into a formal discipline. In his seminal work, *Rhetoric,* he introduced a triadic framework—*ethos* (credibility), *pathos* (emotional appeal), and *logos* (logical argument)—that still underpins every act of influence today.

These concepts were foundational levers engineered to move minds. Ethos refers to credibility—the trustworthiness and moral authority of the speaker. If the audience does not believe you, they will not believe your message. Pathos engages the emotional circuits of the audience, recognizing that humans are not computers evaluating data—they are organisms responding to fear, hope, and desire. Logos, meanwhile, appeals to reason, using structure, evidence, and logic to build persuasive arguments. Together, these form the holy trinity of classical persuasion.

What is striking is how enduring these principles are. From TED Talks to Super Bowl ads, presidential debates to product pitches, ethos, pathos, and logos remain the primary dials we turn to influence thought and behavior. They are embedded in the neural architecture of persuasion.

Following Aristotle, the Roman statesman Cicero expanded the rhetorical toolkit. If Aristotle gave us the blueprint, Cicero gave us the choreography. In his treatise *De Oratore*, he emphasized *decorum*—the idea that a speaker must adapt their tone, style, and content to fit the occasion and audience. A persuasive message must morph to resonate with the moment. Cicero also highlighted *audience adaptation* as a tactical necessity. To persuade does not imply you impose, but tune, adjust your delivery until it resonates with the listener's values, fears, and aspirations. As such, emotional resonance was a strategic imperative.

In Roman society, these rhetorical insights were highly operational. Rhetoric was the operating system of power. Roman education embedded these tools into its very bones. Aspiring leaders were trained extensively in oratory, learning to sway senates, win legal battles, and inspire legions not just with military might but with verbal virtuosity. Rhetoric was as essential to leadership as strategy or swordsmanship.

Consider this: in a society where written communication was limited and mass media were nonexistent; the human voice was the most potent means of transmitting influence. Therefore, the art of persuasion was a survival skill. Those who mastered it could alter laws, mobilize armies, and shape the collective psyche of an empire. The most revered Roman orators, such as Cicero himself, were both admired and feared. Their words could lift or ruin reputations, incite rebellion, or cement peace.

Thus, the legacy of Aristotle and Cicero is the scaffolding upon which modern persuasion is built. In a world saturated with noise, returning to the disciplined art of ethos, pathos, and logos might be the most radical thing a communicator can do.

Middle Ages – Theological and Legal Persuasion

In the medieval world, persuasion traded in its philosopher's robe for something more ecclesiastical and judicial—it moved into churches and courtrooms. Imagine persuasion as a traveling actor: in Ancient Greece, it performed on the democratic stage; in the Middle Ages, it put on holy vestments or a judge's gown and took center stage in institutions that held the power of heaven or land. The stakes were no longer just opinions; instead, they were salvation, reputation, or property.

Let us start with the pulpit. At a time when few people could

Figure 5: Medieval Persuasion

read and nearly everyone believed in hell, the Church became the dominant force in persuasive communication. Preachers were performing emotional and spiritual theater. Their tools were not data or bullet points but vivid stories, symbolic language, and the threat of eternal damnation. These sermons were crafted to target the oldest part of the brain, the survival system, with emotional appeals designed to evoke fear, awe, and guilt.

Imagine a Sunday sermon as a medieval version of a horror story with a moral. Preachers described the torments of hell in gripping detail: fire, brimstone, and demons with pitchforks—designed not to entertain, but to ensure churchgoers obeyed, repented, and gave alms, charitable donations often given to the poor. They were the psychological equivalents of a full-court press. When paired with stained glass windows depicting salvation and judgment, music that stirred the heart, and incense that filled the senses, the entire church experience became an immersive, persuasive environment. The Church, you could say, was the original master of multisensory branding.

Meanwhile, in the legal world, persuasion was being shaped into a more structured form. Courts of law were the other major venue where rhetoric thrived. With the rise of universities and a renewed interest in classical texts, especially those of Aristotle, scholars began to rediscover formal logic and engage in debate. Legal education emphasized the skills of structuring arguments, anticipating counterarguments, and presenting evidence convincingly—skills that remain foundational in modern legal systems.

One standout figure in this intellectual revival was Thomas Aquinas, a Dominican *friar* and theologian. Friars were engaging in society through study, teaching, and persuasive preaching. As a theologian, Aquinas embodied this mission, blending rigorous philosophy with the art of persuasion to defend and explain the faith in ways that continue to resonate today. Aquinas's brilliance was his ability to combine reason and religion. He did not see faith and logic as enemies. Instead, he believed they could work together. In his writings, especially the massive *Summa Theologica*, he structured theological arguments as if he were solving math problems. Question. Answer. Objection. Rebuttal. Each belief was laid out clearly, using a logical structure borrowed straight from Aristotle. He turned theology into a form of debate.

How was persuasion practiced in the Middle Ages? It was more than through speeches alone. It was a cultural system. In churches, persuasion was often conveyed through sermons, symbols, and rituals that moved the heart and frightened the soul. In courts and universities, it emerged as a form of logic, argumentation, and legal precision. Despite the heavy influence of dogma and hierarchy, persuasion still thrived—just in new forms. Whether seeking to save souls or settle land disputes, medieval persuaders proved that influence, when well-crafted, could be as powerful as the sword—and far more enduring.

17th–18th Century – Enlightenment Rationalism

As Europe emerged from the fog of medieval dogma and divine-right monarchies, a new intellectual sunrise began to break the horizon: the Enlightenment. From the salons of Paris to the lecture halls of Edinburgh, thinkers began to ask a dangerous question: What if reason—not tradition, not scripture, not fear—should govern human affairs? With that, persuasion took on a new identity. No longer cloaked in incense or enshrined in Latin liturgy, it now wore powdered wigs and wielded syllogisms.

Figure 6: Enlightenment Rationalism

Philosophers like René Descartes, John Locke, and David Hume led the charge. Descartes famously declared, *"Cogito, ergo sum"* ("I think, therefore I am"), placing rational thought at the very core of human existence. Locke argued that the mind was a blank slate, a *tabula rasa*, shaped not by divine preordination but by sensory experience and education. If people could be reasoned into knowledge, they could also be reasoned into new political systems, ethical frameworks, and scientific truths.

This ushered in an age of *rationalist persuasion,* where the most effective way to convince was not through emotion, authority, or fear, but through logic, clarity, and evidence. Pamphlets replaced pulpits. Argumentation replaced revelation. This was an era of political treatises, scientific manifestos, and constitution drafting as a form of rhetorical art. From Newton's *Principia* to the American Declaration of Independence, persuasion became a printed force, reshaping society not through sword or sermon, but through reasoned text.

And yet, this rational tide was not without undercurrents. David Hume, the Scottish philosopher with a flair for intellectual mischief, warned that reason was "the slave of the passions." He observed that people might justify their actions with logic, but their motivations were still often emotional or instinctual. It was a quietly subversive idea: even in an age obsessed with logic, the heart still pulled the strings.

Still, Hume's insight was a whisper against the roaring optimism of Enlightenment thinkers. Most philosophers of the period believed that if a compelling argument were presented—structured, supported by evidence, and logically sequenced—people would change their minds. This belief, though noble, carried a touch of naiveté. It overlooked the deeper layers of human cognition: the biases, instincts, and emotional currents that truly shape decision-making.

Nevertheless, the Enlightenment laid critical foundations. It gave us the language of rights, the architecture of liberal democracies, and the methodology of science. It believed in persuasion, not as manipulation, but as mutual ascent toward truth. This was persuasion as civil discourse, a meeting of minds rather than a contest of wills. The era gave rise to enduring principles: that ideas should be debated, not decreed; that authority should be earned, not assumed; and that clarity is a persuasive weapon.

Ironically, while Enlightenment thinkers were busy deifying reason, they also planted the seeds for future revolutions—both intellectual and political—that would prove just how unruly and emotional human beings could be. Reason opened the door. Passion walked in behind it.

Late 19th–Early 20th Century – The Rise of Advertising

As the gears of industrialization began to whir and clank across Europe and America, they did not just churn out textiles and automobiles; they gave birth to a new necessity: the art of selling. With mass production came mass surplus, and with it, a new economic problem. Making goods was no longer enough; one had to move them, in mass, into the hands of a skeptical, distracted, and increasingly urban population. Enter the golden age of the early advertising field that evolved not out of poetry or politics, but necessity.

One of the first pioneers of this emerging craft was Claude Hopkins, a man who treated advertising not as an art form, but as a science. In his landmark work *Scientific Advertising* (1923), Hopkins argued that ads should be testable, measurable, and accountable. He was a split-tester long before A/B testing had a name. He introduced coupons and sampling campaigns to quantify response, treating every

Figure 7: The Rise of Advertising

advertisement like an experiment. To him, persuasion was a puzzle that could be solved with enough data. If the ad did not sell, it was not creative; it was broken.

But where Hopkins brought rigor, another figure brought revelation. Walter Dill Scott, a psychologist turned advertiser, introduced an idea that would forever change how we understand the human buyer: logic was not king. It rarely even showed up at the party. In his 1903 book, *The Psychology of Advertising in Theory and Practice*, Scott proposed a radical notion—that the average consumer was not deliberating like a philosopher but reacting like an animal. Decisions were not the outcome of syllogisms, but the product of suggestion, emotion, and habit.

Scott believed that human suggestibility, the tendency to be influenced without much conscious thought, was the hidden driver behind most consumer behavior. Moreover, unlike Hopkins, who measured what worked, Scott sought to understand *why* it worked. He emphasized the power of repetition in forming memory, the use of visual imagery in capturing attention, and emotional resonance in overcoming resistance. In doing so, he laid the groundwork for what we now call neuromarketing—decades before fMRI machines and eye-tracking heatmaps would confirm his intuition.

He also challenged the notion that persuasion had to appeal to reason. On the contrary, Scott argued, effective ads often worked *because* they bypassed logic—an image of a mother comforting a child sold soap better than any list of antiseptic properties. A dramatic headline or a tear-jerking story could move more product than the most comprehensive product spec sheet. In short, Scott was mapping the limbic brain before neuroscience had even given it a name.

Together, Hopkins and Scott represented the twin engines of modern advertising: analytic and emotional, measurable and intuitive—one believed in the power of statistics, the other in the power of suggestion. Moreover, while their methods differed, both were united by a seismic insight: persuasion could be engineered.

This period marked a shift in advertising from hunches, hypotheses, and heuristics. It was no longer about crafting catchy slogans—it was about understanding the architecture of influence. In doing so, early advertisers did not just sell soap or biscuits; they sold the very idea that human behavior could be decoded, predicted, and shaped.

1920s–1930s – Persuasion Meets Mass Media

The interwar years marked a seismic shift in the evolution of persuasive communication, catalyzed by the rapid rise of mass media technologies, including radio and cinema. These new channels offered an unprecedented opportunity to scale messages across populations and cultures—creating not just audiences, but publics.

At the center of this transformation stood Edward Bernays, often hailed as the father of public relations. A nephew of Sigmund Freud,

Bernays blended psychoanalytic theory with emerging media to craft campaigns that targeted not the rational mind, but the subconscious—the primal and emotional engines of decision-making. Rather than merely inform, Bernays sought to influence by associating products and policies with identity, freedom, and emotion. One of his most iconic campaigns, the 1929 "Torches of Freedom" march, reframed cigarettes as a symbol of female emancipation (Bernays, 1928). He orchestrated a stunt in which debutantes marched down Fifth Avenue puffing on Lucky Strikes during the Easter Parade. The imagery was cinematic, the message subversive: smoking was no longer taboo for women—it was a political act.

Bernays was not selling tobacco; he was selling transformation, autonomy, and defiance.

Figure 8: Luky Strikes Campaign

This clever framing used the primal brain's appetite for identity, power, and rebellion to override rational health concerns (Morin, 2023).

This era also witnessed the academic formalization of propaganda studies, particularly in the aftermath of World War I. Governments had realized the power of coordinated messaging to mold public opinion during wartime. Woodrow Wilson's Committee on Public Information, spearheaded by journalist George Creel, had already demonstrated how images, slogans, and narratives could mobilize an entire nation. Inspired by such success, scholars began analyzing how mass communication could be used to manipulate consent—often under the veil of patriotism or progress.

In Europe, the rise of authoritarian regimes further underscored the manipulative potential of propaganda. Joseph Goebbels, the Nazi minister of propaganda, meticulously engineered speeches, films, and radio broadcasts to embed ideology in the German psyche. In the Soviet Union, agitprop art combined political messaging with theatrical spectacle to align citizens with communist ideals. While ethically alarming, these efforts were scientifically instructive: they confirmed that mass persuasion did not require rational assent—it required emotional submission, repetition, and spectacle.

These developments collectively shifted persuasion from the domain of rhetoric to the laboratory of mass influence. The dual-process model of human decision-making—fast, emotional System 1 versus slow, deliberative System 2—was already in operation in practice, even if it would not be formally articulated for decades (Kahneman, 2011). Bernays intuitively understood that the fastest way to shift public behavior was to bypass the neocortex and stimulate the primal brain through imagery, symbolism, and fear—or hope (Bernays, 1928).

The 1920s and 1930s laid the groundwork for the modern age of persuasion. With media as the megaphone and emotion as the accelerant, this period proved that the architecture of influence no longer required reason—it required resonance. The brain did not need to be convinced. It needed to be *moved.*

1940s–1950s – Experimental Psychology

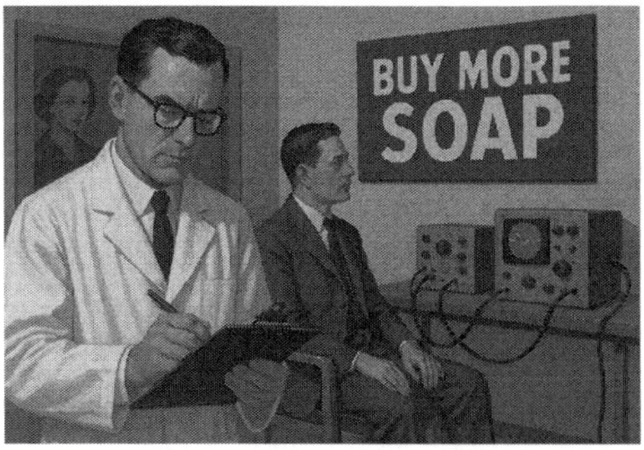

Figure 9: Experimental Psychology

In the wake of World War II, the science of persuasion shifted from intuition and ideology to experimentation and empirical rigor. This was the era when psychology donned a lab coat and got to work decoding what makes messages stick. At the center of this movement stood

Carl Hovland, a Yale psychologist whose research would quietly lay the bedrock for every persuasion strategy used in advertising, political speech, and propaganda to this day (Hovland, Janis, & Kelley, 1953).

Hovland's work was not born in a vacuum; it was forged in the crucible of war. During World War II, the U.S. military faced the enormous challenge of shaping the attitudes and behaviors of millions of soldiers, many of whom had been uprooted from civilian life and thrust into unfamiliar roles. Commanders needed to sustain morale, build unity among diverse recruits, and minimize resistance to challenging or hazardous assignments. Could the right kind of communication make soldiers more receptive and cooperative? Could persuasion serve as a tool of national defense? These urgent questions led the War Department to fund large-scale studies, which became known as the "Yale Communication and Attitude Change Program."

Hovland and his colleagues approached persuasion with a scientific precision that had rarely been seen before. Instead of assuming what "should" work, they ran carefully controlled experiments. Soldiers were shown films, listened to speeches, or read materials, and their attitudes were measured before and after exposure. In doing so, the researchers broke down persuasion into its constituent parts. They asked, 'Does it matter who says it?' How should the message be structured? Which emotions are most effective? When is the best time to deliver it?

One of Hovland's most influential findings concerned the credibility of sources. The team discovered that people were far more likely to be persuaded when the communicator was perceived as both trustworthy and competent. A general, for example, might carry more weight on military tactics than a civilian official, just as a medical doctor's authority could influence acceptance of health advice. This insight foreshadowed the enduring importance of spokespersons, celebrity endorsements, and the cultivation of "brand authority" in marketing.

Another set of studies examined the effects of message structure and order. Should communicators present both sides of an argument or only one? Hovland found that acknowledging opposing views—if done carefully—often increased the persuasiveness of the message, because it made the speaker appear fair and credible. Similarly, he investigated the sequence in which information was delivered, laying the groundwork for what psychologists now call primacy and recency effects: the tendency for people to remember the first and last parts of a message more vividly than the middle.

Emotions, too, came under scientific scrutiny. Hovland examined the role of fear appeals, asking whether frightening messages motivated people to change their attitudes or whether they produced resistance. His nuanced findings suggested that moderate levels of fear, coupled with clear instructions for action, could be persuasive—but overwhelming fear without guidance often backfired. This principle remains in effect in everything from public health campaigns to political advertisements.

Perhaps most groundbreaking was the shift in mindset that Hovland's program represented. Before this era, persuasion had been dominated by rhetoric, philosophy, or ideology. Aristotle's ethos, pathos, and logos provided a conceptual map, but Hovland supplied the scientific method. He transformed persuasion into experimental science, one in which hypotheses could be tested, data collected, and results replicated. This transition mirrored psychology's broader movement from speculation to empiricism in the mid-20th century.

The ripple effects of Hovland's work extended far beyond the military. His findings directly informed the rise of modern advertising, as companies sought to understand consumer attitudes and how to shape them. Political strategists also drew from the Yale studies, recognizing that the credibility of a candidate, the framing of an argument, and the emotional undertone of a campaign speech could decisively sway voters. Even the darker arts of propaganda and psychological operations relied heavily on these insights, applying scientific principles of persuasion to influence large populations.

Hovland's legacy endures in today's communication models. Many contemporary frameworks—such as the Elaboration Likelihood Model or the Heuristic-Systematic Model—owe their lineage to his early experiments. His work demonstrated that persuasion is not a matter of intuition or charisma alone but a science of variables, conditions, and effects. By reducing persuasion to testable components, he made it possible to predict, at least in part, how and why messages succeed or fail.

In sum, the postwar period marked a decisive turning point in the study of persuasion. Thanks to Carl Hovland and his colleagues, the field moved from the philosopher's lectern and the propagandist's podium into the controlled environment of the laboratory. The insights born in that moment still guide the crafting of messages in boardrooms, classrooms, and political campaigns. To this day, every advertisement, speech, or social media post carries echoes of the experiments first conducted in those Yale labs—quiet reminders that persuasion, when understood scientifically, becomes one of the most powerful forces shaping human thought and behavior.

1960s–1970s – Rise of Cognitive Models

By the time the 1950s closed, the blueprint was clear: if you want to change minds, measure first. Then craft. The Hovland era solidified the notion that persuasion was not magic, but rather a matter of mechanics. Moreover, once those gears were exposed, a whole new generation of cognitive scientists, marketers, and behavioral economists was poised to refine them. The persuasive arts had finally become the behavioral sciences.

Thus, the 1960s ushered in a tectonic shift in the science of persuasion. Behavioral psychology, which had long focused on observable stimuli and responses, was now being challenged by a new paradigm—cognitive psychology. This school of thought moved the

Figure 10: Cognitive Models of Persuasion

spotlight inward, emphasizing mental processes such as perception, memory, comprehension, and belief systems as the actual levers of influence.

Central to this cognitive turn was Leon Festinger's very influential *theory of cognitive dissonance* (1957). Festinger proposed that individuals experience psychological discomfort when they hold contradictory beliefs or when their actions conflict with their self-image. This dissonance drives a powerful need to restore internal consistency. For instance, a smoker aware of the health risks may resolve this conflict by downplaying the dangers or switching to a "healthier" brand—a tactic that marketers have quickly exploited. Persuasive messages that align with existing beliefs or offer pathways to reduce dissonance are thus more likely to be accepted.

Cognitive dissonance theory marked a turning point in persuasion research by shifting attention away from external stimuli alone and toward the complex, self-regulating mechanisms of the human mind. It revealed that people are not passive recipients of information but active meaning-makers who strive to maintain harmony between thoughts, values, and behaviors. This insight remains incredibly potent in today's digital persuasion landscape, where conflicting beliefs abound—

Figure 11: Cognitive Dissonance Theory

about products, identities, or brands. In a world saturated with information and cognitive overload, messages that tactfully resolve dissonance can have a profound influence on decision-making, loyalty, and emotional resonance.

While Festinger focused on internal conflicts, William McGuire tackled external threats to belief systems with his Inoculation Theory (1961). Drawing an analogy to medical vaccines, McGuire argued that exposing individuals to weak counterarguments and helping them refute these "cognitive viruses" could build resistance against future persuasive attacks. Imagine a teenager who has been warned about smoking.

Instead of just saying "smoking is bad," a health teacher prepares the student with a weaker opposing message: "Some people might say smoking makes you look cool." Then the teacher helps the student actively refute it: "But the truth is, smoking damages your lungs, yellows your teeth, and makes you smell bad. Is that cool?"

By walking through this process, the student mentally rehearses rejecting the counterargument. So, when a peer later says, "Come on, just try one," the teen is already inoculated—psychologically prepared to resist the pressure, because they have faced and dismissed the logic behind it before. This same strategy is employed in political campaigns, where parties preemptively rebut anticipated attacks, and in brand marketing, where companies address potential criticisms before competitors can leverage them.

This preemptive defense mechanism became crucial in political messaging, health campaigns, and brand loyalty strategies. Just as a vaccinated immune system is better prepared for real pathogens, a mentally inoculated audience is less likely to be swayed by contradictory messages.

Cognitive psychology also emphasized the role of attention, comprehension, and memory in determining persuasive effectiveness. Researchers began examining how message complexity, emotional tone, and structural organization influenced not just whether a message was seen, but whether it was *processed.* This led to refined communication strategies where headlines, layout, and visual anchors were optimized to reduce cognitive load and increase retention. Messages had to navigate the "filtering system" of the mind—something behaviorism had overlooked.

Memory, too, was reimagined not as a passive storehouse but as an active, reconstructive process. This insight transformed advertising and public messaging: slogans, jingles, and repetition were designed to embed ideas deeply into long-term memory networks. The brain was no longer a black box—it was a pattern detector, constantly updating its internal models based on new stimuli and prior knowledge.

These developments collectively shifted persuasion from brute-force repetition toward a more nuanced understanding of belief formation and resistance. It was now clear: the audience was not a blank slate. They were cognitive agents—evaluating, distorting, rejecting, or accepting information based on mental schemas shaped by past experiences.

In short, the 1960s and 1970s turned persuasion into a cognitive chess game. Victory did not go to the loudest message, but to the one that anticipated mental counter-moves and aligned with how the mind processed contradiction, coherence, and defense.

1980s – Dual-Route Models

The 1980s marked a critical conceptual leap in the science of influence: the recognition that persuasion does not follow a single,

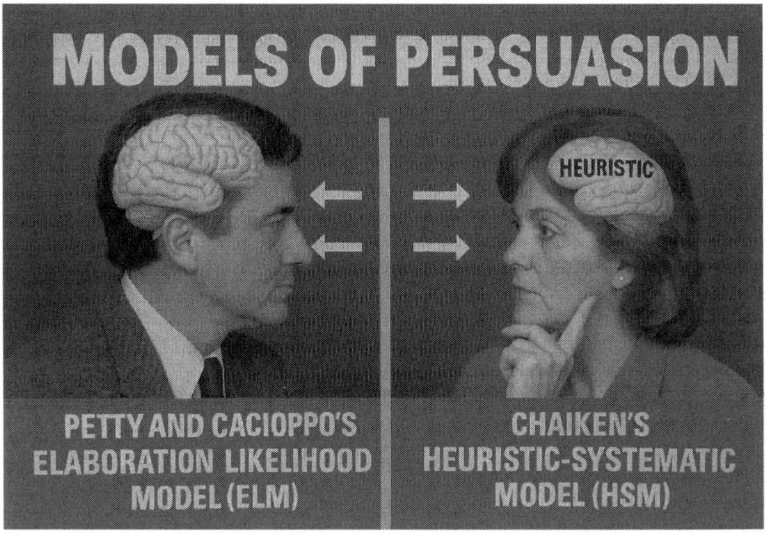

Figure 12: The Dual Route Model

uniform pathway in the brain. Instead, the mind toggles between different processing modes depending on motivation, ability, and context. Two seminal frameworks emerged—Petty and Cacioppo's Elaboration Likelihood Model (ELM) and Chaiken's Heuristic-

Systematic Model (HSM)—which together transformed our understanding of how people are persuaded and why.

Richard E. Petty and John T. Cacioppo introduced *the Elaboration Likelihood Model* in 1986, arguing that persuasion occurs through two distinct routes: *the central route and the peripheral route* (Petty & Cacioppo, 1986). The central route is effortful, logical, and data-driven. It engages when the audience is motivated and capable of scrutinizing the message, like a policy analyst reviewing a white paper. In contrast, the peripheral route is intuitive, fast, and affect-driven. It is the path of least resistance, relying on cues like attractiveness, celebrity endorsements, or sheer repetition. This duality explains why both a well-reasoned essay and a catchy jingle can persuade, albeit through entirely different neural pathways.

Shelly Chaiken's Heuristic-Systematic Model (HSM), proposed around the same time, echoed similar themes but emphasized the *cognitive shortcuts—or heuristics*—that people use to make decisions under uncertainty or low motivation (Chaiken, 1980). These mental shortcuts, such as "experts are usually right" or "if many believe it, it must be true," allow individuals to make quick judgments without intensive cognitive effort. HSM underscored how persuasion could hinge not on content, but on context: the source's authority, the perceived consensus, or even superficial features like color and font.

Together, these models provided a scientific rationale for the success of irrational advertising tactics. Why do we trust a toothpaste simply because it is "recommended by 9 out of 10 dentists"? That is the heuristic "social proof" at work. Why are we drawn to a medication ad because it features a smiling actor in a white coat? That is authority bias operating via the peripheral route. These insights demystified the power of aesthetic design, celebrity branding, and visual packaging in persuasion science.

What made these models revolutionary was their flexibility. They did not just tell us how people *should* process messages; they explained how people *actually* do. The likelihood of engaging in the central route depends on both the audience's motivation (e.g., personal relevance) and ability (e.g., cognitive capacity, lack of distraction). When either is low, the brain defaults to heuristics.

Importantly, these models foreshadowed much of what neuroscience would later confirm: that the brain conserves energy by relying on rapid, emotionally colored judgments unless given a compelling reason to slow down. This dovetailed with Kahneman's later System 1/System 2 framework, reinforcing the dual-process nature of cognition.

By the close of the 1980s, persuasion was no longer seen as a blunt-force operation. It had become a dance between effortful logic and intuitive shortcuts, where success depended less on the strength of the message and more on the path it traveled through the brain.

1990s – Behavioral Economics and Cognitive Biases

The 1990s marked the golden age of behavioral economics, the intellectual lovechild of psychology and classical economics. At its helm were psychologists Daniel Kahneman and Amos Tversky, whose groundbreaking work exposed a startling truth: humans are not the

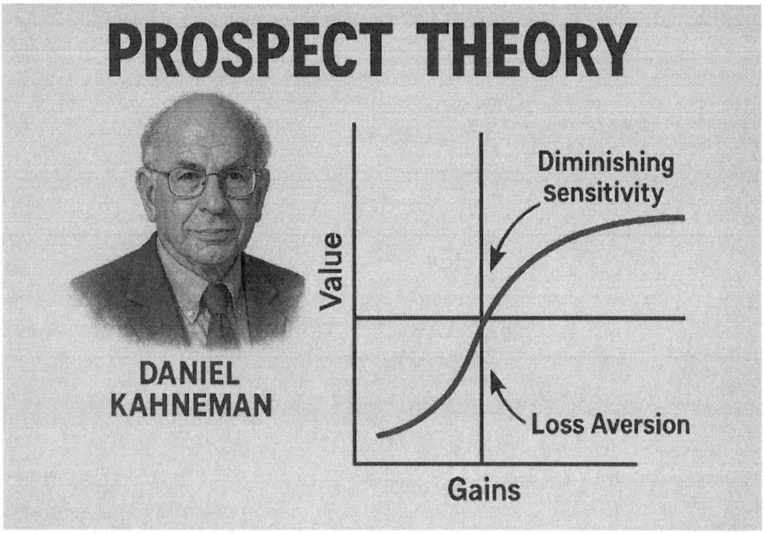

Figure 13: Prospect Theory

rational calculators that traditional economic theory imagined. Instead, we are wired with cognitive biases—systematic mental shortcuts that often lead us to make predictably irrational decisions.

One of their most influential contributions was *the prospect theory*, introduced in 1979 but widely adopted and expanded upon during the 1990s (Kahneman & Tversky, 1979). This theory challenged the idea that people evaluate outcomes based solely on final states. Instead, we frame choices relative to a reference point—typically the status quo—

and weigh potential losses more heavily than equivalent gains. This phenomenon, known as loss aversion, reveals that the pain of losing $100 is psychologically stronger than the pleasure of gaining $100.

Another bias, anchoring, demonstrated how irrelevant numerical cues could heavily skew our decisions. For example, if you are told that the average wine costs $100, you are likely to overvalue a $50 bottle—even if it is objectively overpriced. This simple manipulation of expectations and reference points quickly became a toolkit for marketers, negotiators, and UX designers. Pricing strategies began to use "decoy" options to steer consumer choices and limited-time offers leveraged urgency and loss aversion to close sales.

These insights reshaped the practice of persuasion across industries. Framing effects, a central feature of behavioral economics, demonstrate that the same information can lead to drastically different outcomes depending on how it is presented. For instance, consumers were more likely to choose a surgery when told it had a "90% survival rate" than when told it had a "10% mortality rate"—despite the identical statistics (Tversky & Kahneman, 1981). By the 1990s, such findings had migrated from academia to advertising boards and product design teams.

Kahneman and Tversky's work also laid the foundation for choice architecture, the strategic design of environments that guide decision-making. Whether arranging cafeteria food to encourage healthy eating or creating default options for organ donation, behavioral economists learned to "nudge" people without restricting freedom—a concept popularized later by Richard Thaler and Cass Sunstein (Thaler & Sunstein, 2008).

The 1990s fused psychological realism with economic modeling, replacing the mythical Homo economicus with a more neurologically plausible decision-maker. Marketers, armed with these insights, moved beyond demographics and into the architecture of attention, emotion, and bias. No longer was persuasion merely about logic or appeal—it was about shaping the invisible mental scaffolding behind every decision.

The era proved a pivotal juncture: persuasion had graduated from persuasion *tactics* to persuasion *systems*. By decoding the predictable quirks of the human mind, behavioral economics offered a map of cognitive terrain—and marketers quickly learned how to travel it with precision.

2011 – Kahneman's Thinking, Fast and Slow

In late 2011, Daniel Kahneman released *Thinking, Fast and Slow*. This seminal book would become a cornerstone of modern behavioral science and the most influential book of my entire neuromarketing career. Synthesizing decades of research with his late collaborator Amos Tversky, Kahneman presented a deceptively simple yet profoundly influential model of the mind: *the dual-system framework*. This model forever altered how we understand decision-

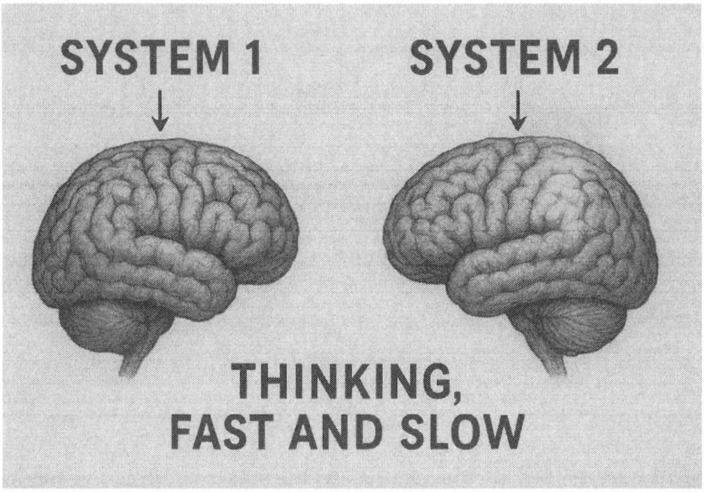

Figure 14: System 1 and System 2

making, judgment, and persuasion. Curiously, Patrick and I proposed a dual brain persuasion theory that maps exceptionally well with Kahneman's book nine years earlier.

At its core, Kahneman's theory divides cognition into two systems. System 1 is fast, intuitive, automatic, and emotional. It governs snap judgments, gut reactions, and heuristics—decisions made without conscious effort. System 2, in contrast, is slow, deliberate, analytical, and logical. It is invoked for complex calculations, careful planning, and reflective thinking (Kahneman, 2011). The book's core message? System 1 dominates. Most of our decisions are not reasoned; they are reflexive.

This insight did more than question previous psychological theories—it revolutionized how professionals across disciplines would approach persuasion for decades. While I had already recognized that most purchase decisions were driven not by careful deliberation, but by emotional cues and mental shortcuts, I also noticed that other marketers began to craft messages that aligned with System 1's preference for speed and simplicity, thereby avoiding cognitive overload. Economists also started to reconsider the rational-agent model, acknowledging the messy and biased nature of real-world choices.

Most importantly, Kahneman's work gave professionals a shared vocabulary to diagnose and discuss cognitive errors. Terms like "System 1 bias," "cognitive ease," and "availability heuristic" entered boardrooms, classrooms, and design meetings. No longer was "irrationality" a moral failing; it was a feature of the brain's architecture.

The book's influence reached far beyond psychology. In finance, Kahneman's exploration of overconfidence, loss aversion, and risk perception offered critical insights into market behavior. In public policy, the dual-system model informed the creation of "nudge units"—government teams tasked with designing interventions based on behavioral science. In education, teachers applied his ideas to foster metacognitive awareness in students.

Critically, *Thinking, Fast and Slow* also served as a mirror. It forced decision-makers—from CEOs to clinicians—to reflect on their vulnerabilities to bias. It challenged the myth of rational objectivity and replaced it with a humbler and more realistic model of human cognition. Given the profound importance of this book to the evolution and growth of the neuromarketing field, we will continue to explore its teaching as we progress through the book.

When Marketing Meets Neuroscience

As the 21st century unfolded, the study of persuasion entered a new epoch—one defined not by surveys or focus groups, but by the direct observation of the human brain in action. Other frustrated marketers and disillusioned academics, along with myself, started to investigate the neurological basis of persuasion, specifically the direct effect of advertising on the brain. Enabled by cutting-edge technologies like functional Magnetic Resonance Imaging (fMRI), Electroencephalography (EEG), and biometric sensors, we no longer had to infer what people thought or felt—we could watch the mind at work in real time. This marked the dawn of neuromarketing, a field that my partner, Patrick Renvoise, and I are credited with pioneering as early as 2002.

Figure 15: When Marketing Meets Neuroscience

At the theoretical level, Kahneman's prospect theory, as well as the triune brain model, first proposed by Paul MacLean, rapidly gained new popularity as a teaching and diagnostic tool to support the main arguments behind neuromarketing models. While not anatomically literal, McLean's model continued to offer a compelling metaphor for understanding consumer behavior as a neurobiological process influenced by the anatomical layering of the brain, from the primal to the rational. In fact, my business partner and I titled our first book "Selling to the Old Brain" (Morin & Renvoise, 2003), using this grabber line as a provocative way to suggest that we had discovered the new frontier of consumer research.

In it, we introduced a practical framework grounded in brain science and case studies—rather than lengthy academic validation. With boldness (and perhaps a dash of that trademark French arrogance), we argued that most decisions are made in the "old brain," also referred to as the reptilian brain—a primitive structure responsible for survival instincts, impulses, and raw emotions. We also identified six message stimuli that directly activate this ancient neural gatekeeper.

First, *self-centered* messaging—because the primal brain is wired for survival, not altruism, and instinctively asks, "What is in it for me?" *Contrast* follows, as the old brain is attuned to differences that signal change, danger, or opportunity; it responds to choices framed in stark relief. *Tangibility* matters because abstract ideas drift past unnoticed, while concrete, sensory-rich information grabs its attention. Then comes *primacy and recency*—the brain retains what it hears first and last, treating these moments as cognitive bookends. *Visual cues* are essential, as vision dominates our sensory processing, and the primal brain reacts more quickly to images than to words. Finally, *emotional resonance*—the universal key—because nothing stamps memory or drives behavior more forcefully than feelings, particularly those tied to fear, joy, or surprise.

We printed fifty copies at Kinko's, then the unofficial mecca for scrappy entrepreneurs—those brave enough to believe that a spiral-bound booklet could launch a movement. We were convinced that all we needed was a book in hand—the ultimate prop—to be taken seriously. And so, armed with paper and persuasion, we invited ourselves to every event, panel, and stage that would have us determined to get people excited about our message.

One moment still burns brightly in my memory: an event organized by the French Chamber of Commerce in San Francisco in 2002. Patrick and I did not come to lecture: we came to perform. Our presentation was meticulously engineered to activate all six primal stimuli. We seized the audience's attention within seconds, kept them off-balance with unpredictable twists and narrative turns, and laced the entire delivery with emotion and visual impact. It was a full-brain experience. And it worked. The response was electric. That night, and at countless events that followed, we encountered the same reaction—astonishment, curiosity, and genuine enthusiasm. Often, to our bewilderment, we realized: we were launched.

Within just a few years, with the help of a publisher, we completely rewrote the text. We published an updated version of our persuasion model under the title "Neuromarketing: Understanding the Buy Buttons in Your Customer's Brain" (Morin & Renvoise, 2007). As we began to revise and strengthen our arguments, several prominent neuroscientists confirmed the critical importance of the limbic system in all our decisions. Neuroscientist Antonio Damasio presented this elegant argument for the primacy of emotion in all our decisions. In his widely cited case study of "Elliot," a man with damage to brain areas responsible for emotion, Damasio revealed a striking truth: without emotion, decision-making collapsed. Elliot could describe choices in detail but was paralyzed when asked to make even mundane decisions, such as scheduling an appointment (Damasio, 1994). This showed that emotion was not the enemy of reason; it was its engine. Therefore, as we understood Damasio's findings, there was compelling evidence that logic required emotional weighting to prioritize, evaluate, and act.

In a swift move, we secured the URL *neuromarketing.com* and positioned our company as a leading training organization in the neuromarketing space, a place we still hold today. Given my passion for research, though, we quickly started to conduct studies using brain-based modalities. This did not completely change our initial model; it strengthened our theoretical framework. Building on the six primal triggers of persuasion, it solidified the relevance and critical importance of our four persuasive steps (Diagnose Pain, Differentiate Claims, Demonstrate Gain, and Deliver to the Old Brain). We also revised and improved our seven "impact boosters," including emotional resonance and storytelling— deepening the original model.

Unexpectedly, our first "published" book is still selling well today and is offered in over 12 languages. While it is difficult to track the exact number of books we sold of this title, we estimate it is close to 200,000 copies, a rare achievement among marketing authors. After all, it highlights how to *apply* brain-based principles in copywriting, design, and messaging, effectively opening the door to what became modern neuromarketing.

I admit that I was not convinced it would be so successful. However, this original model helped garner us some attention. While still lacking scientific credibility, it offered a seductive formula to recommend messages that could first trigger primal safety, then engage emotional resonance, and finally present rational justification—mirroring how the brain processes decisions.

As we had hoped, a firm confirmation emerged from our collective investigations: emotional engagement consistently predicted advertising effectiveness better than conscious preference reports. It was not what people said they liked that mattered; it was what stirred their limbic systems. If the campaign failed to trigger emotion, it vanished from memory, bypassing the brain's ancient attentional filters entirely.

As companies, from Coca-Cola to Google, began integrating neuromarketing into campaign development, an innovative marketing research field emerged—one that viewed persuasion not just as an art, but as a testable, visualizable neural process. Unlike traditional marketing methods that rely on self-report—often flawed due to social desirability bias, recall error, or a simple lack of self-awareness—neuromarketing turned to the body's signals for truth. Metrics such as pupil dilation, heart rate variability, skin conductance (GSR), and facial expression coding provided a rich, unconscious layer of data. If a consumer's face twitched, their pupils widened, or their palms began to sweat while viewing a campaign, these involuntary responses were considered more telling than anything they could verbalize.

Realizing that an MBA alone was insufficient to address the profound complexity of studying the effect of ads on the brain, I felt compelled to pursue more in-depth, scientific training. My MBA in Marketing Research had given me good tools, but not the kind of empirical precision needed to decode the neural mechanics behind persuasion. This realization led me to enroll in a PhD program in Media Psychology with a concentration in neuropsychology at Fielding Graduate Institute (now Fielding Graduate University), where I still teach today—a field at the intersection of cognition, emotion, and communication. For my dissertation, I focused on how advertising affects the brains of teenagers, a demographic that is especially vulnerable to persuasive stimuli.

Meanwhile, I was fortunate to be advised by Dr. Paul Zak, a pioneer in neuroeconomics best known for his groundbreaking work on oxytocin, often referred to as the "trust hormone." Our relationship evolved beyond academia, developing into a business partnership that eventually led to the acquisition of our Imotions neuromarketing testing platform. This investment enabled us to conduct neuromarketing studies with complete autonomy. Paul's mentorship was transformative. He challenged me to think like a scientist while staying grounded in real-world application. I owe much of my development as a neuroscientist to his guidance, generosity, and intellectual rigor.

Today, Zak continues to move beyond neuroeconomics and neuromanagement into building predictive, AI-powered neuromarketing platforms. His work with ML-driven forecasting, wearable immersion measurement, and interdisciplinary research leadership demonstrates how AI is already transforming the science of persuasion under his guidance.

Having traced the winding path of persuasion from ancient rhetoric to the digital frontier, we now leave behind theory and tradition to enter the neural theater where every buying decision truly unfolds. The next chapter opens the skull, metaphorically speaking, to reveal the hidden operator behind our preferences, purchases, and gut reactions. This is where persuasion becomes biology. To influence effectively, we must first understand the primal circuitry that decides what gets noticed, trusted, and remembered. Let us step inside the brain behind every buy.

Chapter Summary

Chapter 1 offers a comprehensive historical overview of persuasion, tracing its evolution from classical rhetoric to the neural era of AI. It begins with Aristotle's triad of ethos, pathos, and logos, establishing the foundational elements of persuasive communication. The narrative then moves through key eras: medieval theological and legal persuasion, Enlightenment rationalism, and the rise of psychological and empirical approaches during the industrial and advertising revolutions.

In the 20th century, persuasion matured into science. Figures like Walter Dill Scott introduced psychology into advertising, Edward Bernays leveraged Freudian insights for mass communication, and Carl Hovland dissected the mechanics of message impact during wartime. The chapter highlights the transition from logic-dominated models to emotionally centered frameworks.

With the advent of behavioral economics, Kahneman and Tversky exposed our cognitive biases, leading to the development of dual-process theories of thinking—System 1 (fast, emotional) versus System 2 (slow, rational). These concepts became critical to understanding real-world decision-making.

Finally, the chapter lands in the 21st century, where neuroscience meets persuasion. Neuroimaging and biometric tools have demonstrated that emotions and subconscious responses, rather than self-reported preferences, are more reliable predictors of behavior. The triune brain model (comprising the primal, limbic, and neocortex structures) is introduced as a metaphor to explain these layered processes.

It was during this pivotal shift that my partner, Patrick Renvoise, and I co-founded SalesBrain, the first neuromarketing agency, and helped shape the very language and models now central to the field. Our development of the NeuroMap® framework provided one of the earliest systematic approaches for applying brain science to marketing, turning abstract neural theory into an actionable persuasion strategy.

The central message: while tools and theories have evolved, the human brain's architecture remains fundamentally unchanged. This historical lens reinforces why modern persuasion must respect biological constraints and capitalize on emotional resonance.

A Quick Primal Exercise

Close your eyes and think of the last time you made a quick decision without much deliberation — maybe choosing what to eat, crossing a street, or answering a casual question. Now replay that moment in your mind, but this time, slow it down. Ask yourself: Did my body react first, or did my reasoning mind lead the way? Most likely, your heartbeat, gut feeling, or muscle tension tipped you off before your rational brain had time to justify the decision. This simple reflection shows how much of your daily life is guided by the primal brain's rapid, unconscious evaluations.

Final Thought

Chapter 1 confirms that persuasion begins not in logic but in instinct. Recognizing the primal brain as the gatekeeper of attention and trust reshapes our understanding of influence. From this point forward, the challenge is not to overload the rational mind but to connect meaningfully with the survival-driven circuits that decide what truly matters.

CHAPTER 2: The Brain Behind Every Buy

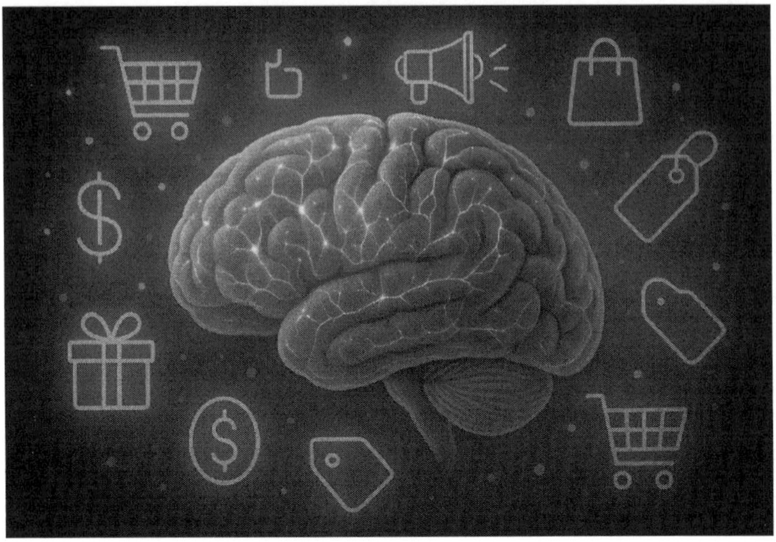

There is no scientific study more vital to man than the study of his own brain. Our entire view of the universe depends on it.

Francis Crick, Ph.D.
Nobel Laureate in Physiology

Before we dive into the science supporting the neuromarketing field, let me take you back to the earlier chapters of my career, long before EEG caps and eye-tracking devices became my daily tools. As the young CEO of Doublet, a global flag manufacturing company in the late 1980s, and later as CMO for Grocery Outlet and ZapMe!, a dotcom in the education and technology space, I wrestled daily with a fundamental question: Why do logical, well-researched marketing strategies often fail to yield results?

Back then, I followed the textbook guidelines of my MBA education—segmenting by demographics, crafting persuasive arguments deeply anchored in logic, and emphasizing value propositions. I instructed my team to craft brochures and pitches based on what people *should* care about. However, I watched in real-time how these carefully crafted messages fizzled. Many campaigns flopped, customers ignored our never-ending lists of compelling reasons to buy our products, and sometimes, even the most data-driven plan backfired. It was not just frustrating, it was humbling.

That frustration became my obsession. I began to suspect that the real obstacle was not in the message itself, but in how people receive and process messages. I finally realized that the concept of the "rational buyer" was a myth. The missing variable was not just age or income—it was the brain.

For Grocery Outlet, primarily, a fast-growing chain of value-focused retail known for offering brand-name groceries at deeply discounted prices through opportunistic buying and a treasure-hunt shopping experience, I had sat through countless focus groups, conducted surveys on thousands of customers, only to find very weak correlations between sales data, direct observations, and self-reported data. Participants told us what they *thought* we wanted to hear—or what made them sound reasonable.

Over time, I came to see the pitfalls of collection methods based on self-reports firsthand. Respondents often suffer from survey fatigue, clicking through questions with minimal thought to reach the end of the questionnaire. And then there are the customers who respond only when they are angry, turning feedback loops into gripe sessions rather than windows into the truth. The emergence of online panels, which gained popularity by offering compensation for survey responses, only exacerbated the problem, producing data that was often superficial at best.

However, this was not a matter of dishonesty all along; it was a matter of biology. Most people lack conscious access to the true forces driving their decisions. "Garbage in, garbage out." I was measuring opinions, not neural responses—and I saw the disconnect with my own eyes. I spent hours watching surveillance footage from my company-owned grocery stores, studying how shoppers behaved. What I witnessed often contradicted the findings of our surveys and focus groups. Customers froze in front of walls of products, paralyzed by the overwhelming number of choices. Others kept their gaze fixed on the floor, scanning for safety as they navigated cluttered aisles. Then, like magic, if a pallet of discounted goods blocked their path, their attention was instantly seized, not because of messaging, but because the environment imposed itself on the senses. These were not abstract findings. These were human behaviors that revealed how the brain responds to its environment in ways that no checkbox could capture. I had been trying to decode the wrong signals. After only two years with Grocery Outlet, I accepted a job as CMO of Zap Me! in 2000, excited to join a technology company in the education space and convinced I would begin to test some of my early assumptions about the psychological basis of consumer behavior.

However, Zap Me! was a surreal chapter in my career. It was a vivid example of how money can turbocharge a tech dream without necessarily grounding it in measurable results. On paper, we had it all: the largest satellite-powered computer lab network in U.S. schools, a slick business plan, and over $100 million raised on the strength of a catchy PowerPoint deck. The startup moved at lightning speed. By the time I joined, about 4 years after the business was created, more than 2,300 schools in 45 states had been wired, providing two million students with internet access via satellite, rather than dial-up. But beneath the surface, a storm was brewing that I had not even seen before I signed up for the job. ZapMe! was not just an edtech darling—it was also a controversial player in the rapidly evolving digital advertising space.

The business model hinged on a tradeoff: free technology for schools in exchange for exposing students to highly targeted banner ads. What sounded visionary in boardrooms (and to me!) triggered alarm bells in classrooms. Critics pounced. Commercial Alert, a watchdog group in D.C., branded us "corporate predators," arguing we were exploiting children's attention for profit. As the political pushback intensified, I found myself at the center of a national debate. I had no choice but to learn—quickly and under fire—how to stand firm when questioned, to speak with clarity when challenged, and, at times, to absorb the criticism without flinching. The scrutiny was not abstract; it came live, in real time, as I sat across from sharp-tongued anchors on CNN and other national media outlets. Interviews often turned into interrogations. But instead of retreating, I learned to articulate complex, ethically charged ideas with conviction and calm.

What I did not fully realize then was that I was being forged for something even more polarizing. The skills I honed—managing controversy, holding my ground in the face of moral ambiguity, and staying mission-focused amid a whirlwind of misunderstanding—would prove essential when I entered the volatile world of neuromarketing. This field would challenge not just what people think, but how they think.

Despite our efforts to monetize the technology and pivot the model, the momentum was not enough. Within two years, we folded. The labs were shut down, patents quietly sold, and nearly 200 team members were sent home. I was stunned. But the universe handed me an unexpected gift: a one-year exit package, a sizeable bonus for closing a $50 million deal—and, most precious of all, the rarest resource in any professional life: uninterrupted time to think, reset, and reinvent.

With that time, and some financial security for the first time in my life, not to mention, in the thick of a divorce, I paused and reflected deeply. I had spent years marketing in traditional industries, such as silk-screen manufacturing and food and beverage distribution. However, I was increasingly convinced that I could do much more by revolutionizing marketing itself. Not by tweaking campaigns, but by rewriting the rules entirely. Not in a legacy sector, but by building a new discipline from the ground up, rooted in what the brain truly responds to.

That was the beginning of everything that followed. I knew then that the new frontier of marketing was not about more segmentation or cleverer taglines—it was about decoding how people process messages in their brains. The path forward was not paved with better questions, but with better measurements of the unconscious mind at work. Shaped by an MBA education that provided only a rudimentary foundation in psychology, I realized I needed to elevate my knowledge. Marketing was not just about numbers and positioning statements—it was about the mind.

So, I returned to academia with renewed purpose and began immersing myself in neuroscience and neuropsychology, eventually earning a PhD in Media Psychology with a focus on neuroscience.

Welcome to the world of the primal brain.

Over the last two decades, I have immersed myself in studying, measuring, and decoding the primal brain to the extent that I am considered one of the foremost experts in the world in explaining and predicting advertising effectiveness using applied neuroscience. From building the first neuromarketing agency to developing the NeuroMap® model used by many Fortune 1000 companies, my partner and I have tested, refined, and taught primal brain strategies in over 30 countries. We estimate that we have trained over 200,000 executives worldwide, maintaining a long-standing strategic relationship with Vistage, the world's largest peer-learning organization. For Vistage, we traveled to hundreds of cities in the US, Canada, Australia, the EU, Brazil, and Southeast Asia, delivering over 3,000 workshops about the teachings of our neuromarketing model. For our achievements. Patrick and I each received multiple speaking awards individually.

From the beginning, this work has felt like a mission, not just a career. It sharpened my perspective on what truly drives human decision-making. It helped me understand, decode, and teach others how to engage our silent and all-mysterious decision-maker within ethical boundaries. The primal brain has been at the center of thousands of experiments I have run, campaigns I have designed, and leaders I have trained. I do not just talk about it—I have staked my entire reputation on it.

Meanwhile, the term "reptilian brain" is often used interchangeably with the term "primal brain," especially in marketing circles. However, it is essential to acknowledge that this label is a subject of controversy among neuroscientists. And rightly so! Critics (and I am one of them, curiously) argue that calling it "reptilian" oversimplifies brain evolution and incorrectly implies that we have a literal reptile brain sitting beneath our more advanced structures. This distinction is significant because it sets the stage for understanding how evolutionary brain structures influence modern behavior, particularly in contexts such as marketing and persuasion. If your message fails to reach this target, you have already lost your audience.

The Triune Brain Model

Briefly introduced earlier, it is worth explaining Paul MacLean's model now since it has remained popular among neuromarketers and scholars. Though most neuroscientists have nuanced (if not debunked) some of its central assumptions, the model remains a potent metaphor for understanding human behavior, especially in the context of persuasion and marketing. As Ploog (2003) clarified, it is less a rigid anatomical

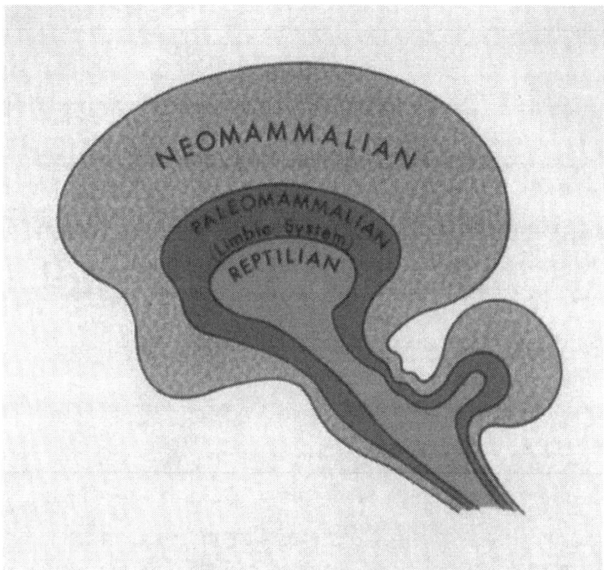

Figure 16: The Triune Model

division and more a functional construct that helps decode how different brain systems influence thought and behavior.

According to McLean, the layers solve distinct, yet complementary priorities that require varying degrees of sophistication and automation.

1. **The reptilian brain (as named by McLean)** is the oldest and most primitive structure. This part of the brain is primarily concerned with survival. It governs functions such as heart rate, breathing, territoriality, and instinctual behaviors. It responds with speed and certainty to perceived threats or rewards. Importantly, it does not "think"—it reacts. From a marketing perspective, this is the brain that

scans for relevance and danger, and only when it detects a potential gain or loss does it permit higher-level processing to occur.

2. **The paleomammalian brain (Limbic System)** emerged next in the evolutionary timeline and introduced the capacity to feel. It governs emotional responses, social bonding, and the encoding of memory. The amygdala, hippocampus, and hypothalamus all play crucial roles in this process. This brain tells us whether we like or dislike something, whether we trust or fear, and whether we are emotionally compelled to act. It interprets the emotional valence of a message, shaping not only how we remember an experience, but also whether we act.

3. **The neomammalian brain** is the crown jewel of human evolution, giving rise to complex thought, language, planning, and reasoning. It allows us to reflect, imagine the future, and weigh abstract concepts. It is slow, energy-intensive, and easily overwhelmed by complexity. This is why we often mistakenly assume that this is where persuasion occurs, involving the crafting of logical arguments, detailed comparisons, and rational explanations. But without the reptilian and emotional green lights, these messages never register.

MacLean's model, while metaphorical, continues to offer enduring explanatory power, particularly for those seeking to influence behavior. Traditional marketing often operates under the illusion that buyers are deliberate, calculating agents. But the triune model dispels that notion. If the reptilian brain, the gatekeeper, does not sense relevance, urgency, or safety, the rest of the brain never gets the message. It says, "This matters," or it says nothing at all.

When I first began working on neuromarketing consulting engagements in the early 2000s, I observed a considerable number of marketers and advertisers making that critical mistake across various industries. One case involved a B2B software firm with an exceptionally detailed product offering in the software as a service (SAAS) space. Their marketing materials read like engineering manuals: logical, exhaustive, and utterly forgettable. Sales were flat. Closing ratios were abysmal. Yet the value proposition was unique. Why the hassle then? Because the message did not stimulate a primal trigger and landed directly in the neocortex—too early, too much to decipher with logic alone. Once we restructured their message to emphasize primal fears, such as wasting time ("launch in minutes"), cognitive effort ("no training required"), and the avoidance of downtime ("best uptime"), their conversion rate jumped nearly 40%.

In another case, a healthcare company had a cutting-edge device that could save time during surgical procedures. Their messaging emphasized data, clinical studies, and graphs—logical appeals. However, when we reimagined the story through the triune lens—leading with the story of a panicked nurse needing to save a life in seconds—the emotional resonance increased significantly. Surgeons not only understood the value; they felt it. Moreover, feeling, as modern neuroscience confirms, precedes action.

We described the triune brain like a security system with a receptionist. The reptilian brain was the security guard, deciding who gets into the building. The limbic system was the lobby, forming impressions and emotions. Only then could you reach the executive suite, the neocortex, where decisions were formalized. If your marketing content does not get past the lobby, it will never reach the boardroom.

This simple model served as our first compass. It reminded our early customers that successful persuasion did not start with what they said; it started with how their messages made their targets feel. The more they aligned with the triune sequence—trigger, connect, explain—the more likely their audience would not only listen, but act.

From the Triune Model to System 1 and System 2

You recall I have already introduced Dr. Daniel Kahneman's Prospect Theory and his landmark book *Thinking, Fast and Slow*. He remains one of the very few social scientists to have been awarded the Nobel Prize in Economic Sciences. This rare recognition underscores how deeply his insights have reshaped not only psychology but also the very fabric of economics and decision science.

But beyond the theories and accolades lies a profound human story. A year ago, Dr. Kahneman made a deeply deliberate and personal choice. Shortly before his 91st birthday, he traveled to Switzerland to end his life through assisted suicide. In a farewell letter to close friends, he wrote:

"This is a goodbye letter… I am on my way to Switzerland, where my life will end on March 27... I have believed since I was a teenager that the miseries and indignities of the last years of life are superfluous… I discovered… I am not afraid of not existing… death is like going to

sleep and not waking up… Thank you for helping make my life a good one."

He was not terminally ill. He remained intellectually sharp. However, he feared the slow erosion of mind and body. At 90, he weighed the diminishing joys of life against its mounting burdens. In a way, it was the ultimate exercise in cognitive clarity—a decision made in complete alignment with the very principles he spent a lifetime exploring.

The world learned of his passing only months later. It came as a jolt. Could the world's leading authority on decision-making truly choose to end his life? Moreover, could that decision be both rational and compassionate? As with so much of his work, Kahneman's final act invites us to think deeply about autonomy, dignity, and what it means to live and die with deliberate intent. His legacy endures not only in the textbooks and theories he authored, but also in the courage he demonstrated in applying those principles to life's most intimate and irreversible decisions, proving that wisdom is not just knowing how we choose, but also knowing when to let go.

Born in Tel Aviv and raised in France during the Nazi occupation, his early experiences with human fear and behavior influenced his academic path. After studying psychology and mathematics, he began his career in the Israeli Defense Forces, where he worked on psychological evaluations for recruits. It was there that he began questioning the accuracy and consistency of human judgment, a curiosity that would eventually shape his life's work.

As noted earlier, Kahneman's groundbreaking collaboration with Amos Tversky in the 1970s and 1980s fundamentally altered our understanding of judgment, decision-making, and risk. Together, they uncovered the many cognitive biases that skew human reasoning, introducing concepts such as loss aversion, anchoring, and availability heuristics—now cornerstones of behavioral economics. These insights culminated in the development of dual-process theory, which posits that the human mind operates with two distinct but interconnected systems.

Published in 2011, Kahneman's book *Thinking, Fast and Slow* became a global bestseller and a foundational text in both psychology and economics. The release of his book provided us with the scientific framework we had been longing for to support the value of NeuroMap®, which had been introduced nearly a decade earlier. Kahneman's dual-process theory outlined a compelling map of how we think: one system that operates automatically and another that engages with effort. It resonated with marketers, educators, and decision scientists alike, offering a language to describe the tug-of-war between instinct and analysis that plays out in our minds daily.

Few people recognize, however, that dual-process theories did not begin with Kahneman and Tversky. The intellectual roots go back over a century, threading through early models of human cognition that we discussed in Chapter 1. Even William James, often referred to as the father of American psychology, distinguished between automatic and effortful attention at the end of the 19th century. Additionally, philosophers such as Henri Bergson emphasized the contrast between intuition and rational deliberation.

Then, of course, it was decades later that the elaboration likelihood model (ELM) proposed by Petty and Cacioppo in the 1980s framed two routes to persuasion: the central route, which requires high cognitive effort and scrutiny (akin to System 2), and the peripheral route, which relies on cues like attractiveness or credibility (closer to System 1). Additionally, Chaiken's heuristic-systematic model (HSM) proposed a dual-processing system, arguing that messages are evaluated through either mental shortcuts or detailed analysis.

Together, these models shared a common insight I could not miss: people do not weigh every decision with equal deliberation. Most judgments are fast, emotionally charged, and heavily context-dependent. Kahneman and Tversky offered empirical clarity and cognitive architecture to these observations. Their language—System 1 and System 2—became shorthand for describing mental operations that had previously been studied in isolation.

With that exciting backdrop, Patrick and I decided to revise our initial model to focus on the dynamic relationship between two brains rather than three layers (Triune model). While the labels System 1 and System 2 were still popular among behavioral economists, we felt they lacked the clarity and specificity of our distinction between primal and rational processing functions. For us, the primal brain is like System 1: fast, automatic, emotional, and often subconscious. Moreover, System 1 decides in milliseconds whether a message is safe, novel, threatening, or rewarding. It is the brain's autopilot. Meanwhile, we consider the rational brain to be similar to System 2. It is slow, deliberate, logical, and conscious. It is the brain's manual override. System 2, meanwhile, is lazy and avoids cognitive effort when confronted with messages that lack primal codes. In that case, it prefers not to be activated. Essentially, if System 1 can resolve a situation independently, System 2 does not even come into play (Kahneman, 2011).

This is how we developed the concept of the bottom-up effect. A message is not processed unless it first ignites the primal brain. Imagine walking into a luxury car showroom. Before the salesperson says a word, your System 1 has already formed impressions: the lighting, the smell of leather, the purr of the engine. You might think you are evaluating specs and financing, but your decision was made when your primal brain whispered, "This makes me feel powerful."

In practice, this means marketers must design stimuli that reach System 1 with immediacy. Visual contrast, emotion, and storytelling quickly capture attention. But once engaged, you must also provide a coherent structure that can satisfy System 2, should it awaken. The best campaigns do both: they spark a gut feeling and withstand rational scrutiny.

But while the parallels between Kahneman's model and our Primal/Rational persuasion framework are self-evident, the models are not identical—and the differences matter. Kahneman's Systems 1 and 2 were designed to describe cognitive operations across a wide range of judgments and decisions, primarily in the context of biases and heuristics. His work is rooted in cognitive psychology and aims to explain the mechanisms behind flawed reasoning and probabilistic error.

The primal and rational model, by contrast, was born out of the need to influence buying decisions, rather than merely analyzing them. It is grounded in evolutionary neuroscience and optimized for communication and persuasion. The "primal brain" is survival-oriented. It is not just heuristic; instead, it is emotionally anchored (including the limbic areas that McLean considered the mammalian layer) and highly responsive to sensory input. In this sense, the primal brain is not just a mode of thinking; it is a sensory-emotional filter that decides—before consciousness kicks in—whether a message is even worthy of further attention.

This is where the concept of the bottom-up effect is so central to our persuasion model. In NeuroMap®, we argue that no message reaches the rational brain unless it first activates the primal brain. Persuasion, therefore, is not a top-down intellectual exercise. It begins in the body—

in the gut—with primal recognition. Only then does the rational system get invited to the conversation.

In short, Kahneman's model explains how we think. Our model explains how to prime people before they have a chance to think.

The Brain is an Energy-Conserving Decision Engine

Why does System 2 or the rational brain stay in the background? Because thinking is metabolically expensive. The human brain accounts for only about 2% of our body weight, yet it consumes roughly 20% of our daily energy (Raichle & Gusnard, 2002). Evolution wired us to conserve mental effort. That is why our brain loves shortcuts—heuristics, biases, and instincts.

From the moment we wake up, this energy-saving mode is in play. Consider your morning routine. You reach for the same mug, brush your teeth in the same order, and follow a consistent path through your house—not because it is efficient from a productivity perspective, but because it is neurologically economical. It avoids using the rational brain, which would require more energy to make fresh decisions.

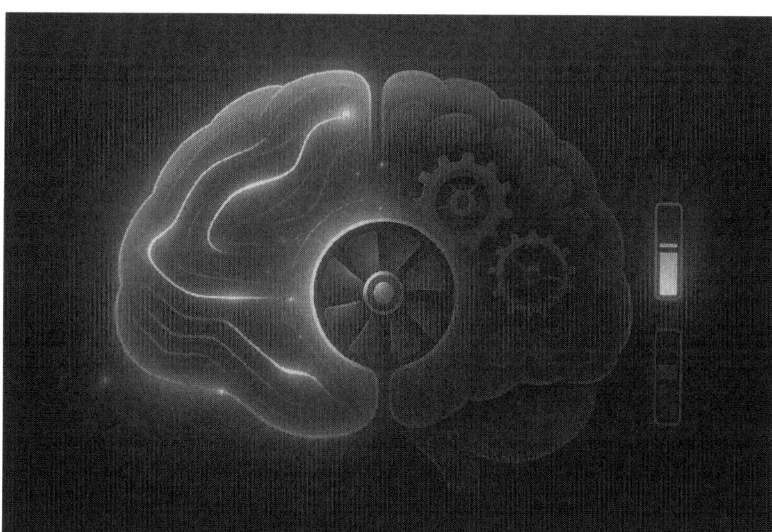

Figure 17: The Brain as an Energy-Conserving Machine

In your commute, do you remember each turn you take? Unlikely. Your brain "offloads" these tasks to the primal brain, allowing you to drive unconsciously. When you suddenly realize you have arrived at your destination with no memory of the route, that is the energy-conserving brain at work.

Decision fatigue is another example. Studies show that as the day progresses, our ability to make logical, thoughtful decisions deteriorates (Baumeister et al., 1998). That is why judges tend to grant parole more often in the morning than in the late afternoon. Your is strategically conserving resources for when they are genuinely needed.

This principle extends to digital behavior. Why do you bounce off websites that make you think? Because every click, every scroll, every menu that is not intuitive is requesting activation of your rational brain. The result? Frustration and disengagement. Friction kills momentum.

Grocery shopping reveals even more. When I watched footage of consumers choosing between 10 brands of the same cereal, they froze because their brains were overloaded. Instead of evaluating each option, they would use heuristics, such as choosing the brand with the brightest packaging or the one they remembered from childhood. These are not rational decisions; they are survival decisions aimed at conserving energy. This has enormous implications for persuasion.

A message that demands cognitive effort is like handing someone a 50-page instruction manual at a cocktail party. Your job as a communicator is to reduce friction, not increase it. Neuroscience-informed messaging must recognize this biological imperative. The less energy your message requires to be understood, the more likely it is to stick and drive action.

This is not a theory I subscribe to; it is one I have tested over the course of two decades. Throughout my career, I have been directly involved in the creation and neurotesting of thousands of persuasive messages—each engineered to penetrate the primal layers of human cognition. These messages have represented hundreds of distinct product lines, spanning everything from nutritional bars to complex biotechnology solutions. I have worked across a broad spectrum of formats, including web pages, billboards, corporate videos, television commercials, pitch decks, print ads, product packaging, keynote presentations, and even commencement speeches.

Whether the goal was to trigger a click, secure a multi-million-dollar deal, or inspire a graduating class, each message was calibrated through neuromarketing frameworks or tools to measure and predict the impact on attention, emotion, and engagement. Some of these campaigns targeted audiences in over 80 countries, offering a rare lens into how the human brain responds to persuasion across cultural, linguistic, and contextual boundaries.

Why Most Ads Fail to Persuade

Advertising now struggles not only to persuade, but to prove its worth. Agencies worldwide are struggling to maintain relevance and profitability, a challenge that is now reaching critical mass with the proliferation of Generative AI. The traditional model—charging a premium for creative directors, copywriters, and big-budget productions is under fire, if not becoming completely obsolete. With AI now generating headlines, video scripts, and ad visuals in seconds, clients are demanding proof of impact. In many cases, the results are not there.

Let us face it; it is not going to get better anytime soon. Agencies that once operated with glossy swagger are now being squeezed by clients who question why they are paying for human intuition when machine learning can A/B test one hundred iterations overnight. According to a 2023 industry survey by Deloitte, over 60% of global marketing leaders expressed concerns about the long-term viability of traditional agency partnerships. Many cited AI's speed, scalability, and data-backed optimization as reasons for rethinking their creative spending.

Meanwhile, creative team members, who were once the stars of the agency world, are now under scrutiny. Their ideas, though often clever and polished, struggle to compete with real-time, neuro-informed, and performance-driven content. The situation worsens when they overlook how the primal brain processes information. Even today, I continue to see expensive campaigns run old tactics from the Mad Men playbook, failing to hit basic metrics for attention and recall.

The bottom line? The challenges facing the industry are not due to a lack of creativity, but rather a failure to evolve. Neuroscience has long confirmed that messages must be perceived before they can be understood. The old approach— "trust us, we know what is clever"—no longer cuts it in a world demanding measurable persuasion.

Meanwhile, advertising remains a trillion-dollar industry, despite a staggering failure rate. According to Nielsen, up to 75% of new product launches fail each year despite massive advertising budgets. A report from the Advertising Research Foundation (2020) found that only 16% of ads deliver a positive return on investment. These numbers are not just bad—they are indictments of a system that often misunderstands or disregards how the brain processes messages.

Take digital advertising, for instance. Billions of dollars are spent annually on banner ads, yet the average click-through rate hovers around 0.1%. That means 999 out of 1,000 people do not even engage. Why? Most ads are cluttered, overly rational, and fail to grab the primal brain's attention. They are designed for the neocortex, not the primal gatekeeper.

Television does not fare much better. A 2017 study by Ipsos showed that 75% of TV ads fail to get noticed, and even fewer are remembered a week later. What about the Super Bowl, the crown jewel of advertising? Despite record-breaking spending, many of its ads are forgotten as soon as the snacks are gone. This is a failure to recognize the power of neuroscience-informed messaging.

As you would suspect, I have had my share of squirm-in-your-chair meetings with CMOs of global giants trying to justify ad budgets so bloated they could bankroll a small country—and still, not a blip on ROI. Picture this: me, introduced as "the neuroscientist" in a boardroom packed with sharply dressed ad agency execs (names withheld for diplomacy and drama). The air is thick, smiles are tight, and you can practically hear the unspoken, "Oh no, this expert (me) is here to read our creative minds—literally."

Clearly, I believe persuasion is more than a parlor trick—it is a high-stakes endeavor where the margin for error can define the outcome of an election, the success or failure of a product, or the reputation of a brand. A striking example unfolded during the most recent U.S. presidential campaign. The Democratic strategy leaned heavily on rational appeals—policy depth, data-driven claims, and logical critiques. While factually rigorous, these messages bypassed the brain's primal gatekeeper, which prioritizes emotional salience, simplicity, and perceived threats, affecting millions of voters. In contrast, the conservative messaging machine embraced short, potent, emotionally charged triggers—focusing on fear, identity, and economic survival. This primal-first approach proved more neurologically aligned with how a majority of voters cast their choice. When persuasion ignores the architecture of the brain, even the most reasoned argument can fall flat. In this game, the cost of failing to move the brain can be colossal.

So, despite decades of behavioral science, many marketers still focus on information rather than emotion. They stack up facts, throw in industry jargon, and proudly display complex diagrams. The result? Messages that bounce off the primal brain like a tennis ball off a tank.

Engaging the Gatekeeper

So, you can think of the primal brain as the gatekeeper to attention. If your message fails here, nothing else matters. That is why the NeuroMap® model, which I co-created with Patrick Renvoise, which we will explore in depth soon, starts by considering the level of pain your products, services, and ideas are designed to eliminate. The brain is more motivated to avoid loss than to achieve gain. A point that Kahneman made extremely clear with the Prospect Theory. Only once the relevant pain threshold is stimulated can a claim be introduced and a gain delivered. A claim is the promise of benefit, but it must follow the triggering of a relevant pain point to engage the brain's decision-making system.

In neurological terms, the primal brain functions as a keeper of peace, a protector of non-urgent requests, and assesses incoming stimuli

Figure 18: The Brain Gatekeeper

in milliseconds to determine whether they are worth the brain's limited cognitive resources. This preconscious evaluation, centered around relevance, urgency, and the promise of relief or gain, determines whether a message is even permitted to reach higher cortical areas for rational deliberation.

Just like a seasoned nightclub bouncer, this gatekeeper does not need a long resume or a complex rubric—it just needs a gut feeling that says, “This is worth it.” We do this countless times a day. We scan emails and decide which to open. We scroll through social media and stop at images that move us. We hear a friend's voice over the phone and intuitively know if something is wrong. These rapid decisions, often made without effort or awareness, are the handiwork of the primal brain evaluating if an “ask” is worth the mental cost.

Thus, to engage this ancient structure (and do so responsibly), you must understand its urgent directive: detecting threats, ensuring safety, and conserving energy. Our primal brain is cautious, ever-vigilant, and biased toward survival. In an evolutionary blink, it scans the environment to assess whether what it is sensing is dangerous, demanding, or delightful.

So, before you even begin to layer in emotional appeal or logic, you must pass this gatekeeper's three-part screening:

- Trigger **relevance** ("This affects me")
- Establish **urgency** ("This affects me now")
- Promise **relief or reward** ("This will make things better")

You have less than 10 seconds to do this. In fact, it is more like 3, if I am honest, because that number keeps decreasing as we face more messages knocking on our brain!

Personal Reflections and Lessons from the Field

Over the course of more than two decades of neuromarketing research and practice, I have encountered nearly every imaginable persuasion mistake. One vivid memory comes from working with a luxury hotel chain whose marketing centered on square footage, bed count, and descriptions of granite countertops. Beautiful amenities, yes—but utterly forgettable. When we retooled their messaging to evoke the feeling of arrival, the sound of a welcoming lobby, and the warmth of human connection, their bookings surged. They had not changed their product—just their appeal to the primal brain.

In another project, we collaborated with a disaster relief organization that was struggling to raise funds for hurricane preparedness campaigns. Their original messaging featured to-do lists and graphs and was entirely focused on "what to do", rather than "why do it?" We introduced stories. Faces. Fear. Hope. The primal brain responds not to pie charts, but to human stakes. Donations and engagement surged within a few days.

These examples are not exceptions. They are patterns. In each of these busy cases, selected from thousands, the primal brain decides, then allows the rest of the brain to catch up and justify it. This profound insight has helped countless leaders—from startup founders to defense contractors—reshape the way they communicate.

Meanwhile, what continues to surprise me is how few marketers are trained to use any theoretical model of persuasion at all. A 2021 meta-analysis by Cornelissen et al. found that marketing campaigns anchored in empirically validated theories, particularly those grounded in behavioral and cognitive neuroscience, outperform those relying on traditional models by more than 30% in consumer engagement and conversion rates (Cornelissen, R., Pandelaere, M., & Warlop, L., 2021). Yet, many still cling to outdated frameworks, such as AIDA (Attention, Interest, Desire, Action) or Maslow's hierarchy, treating them as gospel despite their lack of neuroscientific grounding. These models may hold historical value, but they do not withstand the scrutiny of modern brain research.

In fact, marketers often default to surface-level tactics, mistakenly equating volume with impact. They stack buzzwords, inflate claims, and lean heavily on rational appeals that sound impressive in boardrooms but fail to engage the brain's decision centers. Without a working model rooted in how humans make decisions, their campaigns become acts of creative guesswork. We need better theories, not more theatrics.

My insights are forged through decades of hands-on work, from EEG labs to intense boardroom discussions across over 30 countries. When I co-founded SalesBrain back in 2002, it was not because I wanted to follow trends. It was because I believed persuasion could be elevated to a science—something measurable, testable, and repeatable. I devoted my life to decoding what motivates people to say "yes," not in theory, but in the real world. Whether I am training CEOs, analyzing neuroimaging data, or building a new AI model, the mission remains the same: to make persuasion more innovative, more ethical, and grounded in the way our brains work.

Now that we have explored the architecture and instincts of the brain that drive most of our decisions, it is time to sharpen our map. In the next chapter, we zoom in on the actual functional and neurobiological organism behind the marketing brain. You will meet its two dominant systems: one primal and impulsive, the other rational and reflective. Understanding how these complex systems interact is the marketer's Rosetta Stone for decoding behavior and designing messages that not only inform but also transform.

Chapter Summary

To understand the future of marketing, we must begin with its gatekeeper: the brain. This chapter dismantles the long-held myth that consumers are rational agents who carefully weigh the pros and cons. Instead, we learn that most decisions—many experts argue up to 95%—are made beneath the surface of awareness, in parts of the brain built not for analysis, but for survival. The brain, as this chapter argues, is not a passive recipient of information. It is an energy-efficient prediction engine that constantly scans for threats, rewards, and relevance.

While the triune brain model—dividing the brain into the Reptilian, Limbic, and cortical layers—offers a helpful starting point, it is not sophisticated enough to become the final word in understanding consumer behavior. In this chapter, I introduce a more refined and neurologically grounded revision: a dual-brain system that integrates Kahneman's seminal work on System 1 and System 2 with my own neuromarketing experience.

The Primal brain, which functionally includes both the so-called reptilian structures and the limbic system, is the fast, automatic engine of decision-making. It processes threats and opportunities in milliseconds, long before conscious awareness kicks in. It does not speak the language of features and benefits—it speaks urgency, emotion, and visual salience. This brain does not ask, "Is this product better?" It asks, "Is this safe? Is this rewarding? Is this familiar?" The Limbic system, nested within this primal network, imbues stimuli with emotional significance. It forms memories not just of facts, but of feelings. That is why emotionally charged narratives, faces, and visceral images are remembered long after reasons.

Then comes the Rational brain: slower, deliberative, and effortful. This is the cortex's domain—where we justify decisions, not make them. Kahneman's System 2 resides here, constantly crafting plausible explanations for choices already made by System 1. Most marketing fails because it tries to start here—engaging logic before gaining permission from the fast, emotional brain that governs action. Importantly, I also spotlight a core neurological constraint: the brain is an energy miser. Faced with complex or ambiguous input, it shuts down or stalls. This makes simplicity, clarity, and emotional resonance not stylistic flourishes, but biological imperatives.

So, the real story unfolds when I examine how decisions are made—moment to moment, beneath awareness—through the interplay of two systems with vastly different speeds and priorities. In the next section, we leave behind the anatomical metaphor and step into a functional model that reshapes everything: the Primal/Rational brain system. Rooted in Kahneman's dual-process theory and refined through our neuromarketing lens, this model reveals how persuasion truly works.

A Quick Primal Exercise

Next time you come across an ad, email, or sales pitch, ask yourself: Which part of my brain is this talking to? If it aims for logic before grabbing emotion or attention, it is skipping the gatekeeper. Consider how *you* would reframe it to connect with your primal brain.

Final Thought: Speak Primal First

Imagine the brain as a tri-level building. The primal brain is the doorman. The rational brain is the conference room upstairs. Most brands try to host their persuasive party in the conference room, but the doorman never lets the guests in.

Speak to the primal first. That is how you earn the right to engage emotion, and eventually, logic.

CHAPTER 3: Meet the True Decision Maker

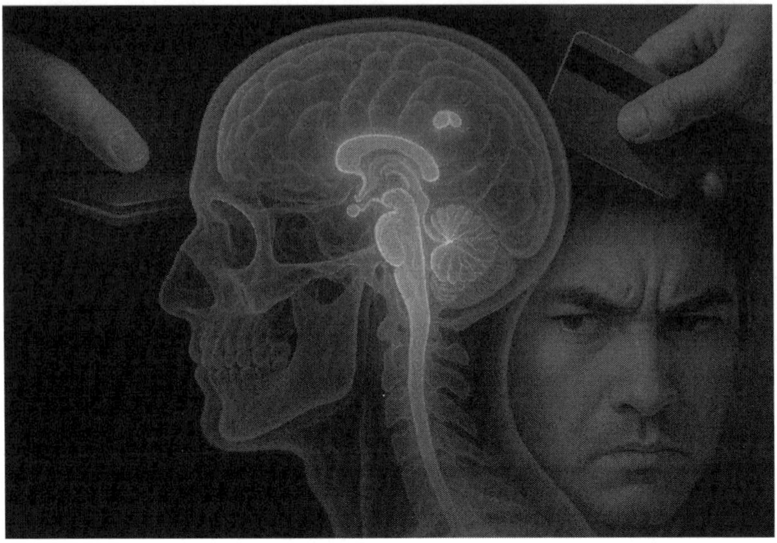

The conscious mind is not at the center of the action in the brain; instead, it is far out on a distant edge, hearing but whispers of the activity.

David Eagleman, Ph.D.

Author of Incognito: The Secret Lives of the Brain

Welcome to your audience's brain. Before you can truly persuade anyone, you need to understand the organ that houses their fears, desires, and, yes, their most instinctive decision filters. For marketers, sales professionals, and communicators, the brain is the final frontier. It is where every message is processed, every decision begins, and every "yes" or "no" takes shape. Long before a product is evaluated or a service compared, it is the biological circuitry of the brain that serves as judge, jury, and executioner of our marketing fate.

And yet, despite its role in governing every click, every hesitation, every wallet pull or brand aversion, the brain remains a black box for most communicators. The typical marketing playbook—crammed with price points, bullet lists, and polished language—barely scratches the surface of what truly motivates a *genuine 'yes'*. If persuasion is the art of turning attention into action, then neuroscience is the blueprint of that transformation.

This chapter is your neuroscience passport. I will take a guided tour of the brain's relevant structures, functions, and quirks, shedding light on why humans often make emotional decisions and justify them rationally. This is not just a clinical lecture. Instead, it is a decoder ring for human persuasion—packed with vivid metaphors, real-world implications, and evolutionary context. Buckle up. You are about to explore the neural terrain where marketing either thrives or fails. Think of it as marketing's operating system: invisible, powerful, and essential for performance.

The Primal and Rational Brain Model

As noted in the conclusion of Chapter 2, our persuasion model builds upon the dual-system architecture popularized by Daniel Kahneman, which is comprised of System 1 and System 2. In Chapter 3, I reframe and augment his theory to fit marketing goals, emphasizing the persuasive roles played by the primal brain (fast, reactive, emotional) and the rational brain (slow, deliberate, logical). These are not isolated anatomical structures but interacting functional systems that reflect how we process influence at distinct levels of urgency and complexity.

To reiterate, while Kahneman's System 1 and System 2 model revolutionized our understanding of decision-making, it was never intended as a roadmap for persuasion. Its brilliance lies in exposing cognitive biases and judgment shortcuts, not in guiding how to engage and influence behavior. That is why I found System 1 to be too broad and blunt an instrument. More importantly, the notion of "fast thinking" reminded me too much of a capability technically powered by the rational brain, without the anatomical or emotional granularity needed to craft compelling messages. That is why the transition from Kahneman's terminology to the Primal/Rational brain framework was not just semantic; it was fundamental.

Thus, Patrick and I refined and expanded our model to reflect both the structure and function of the brain as they relate to the persuasion process. For us, the Primal brain is not just "fast"—it is visual, emotional, vigilant, and evolutionarily tasked with survival. Likewise, the Rational brain is not simply "slow"—it is the post-rationalizing narrator, often playing catch-up to decisions already made. This reframing provides marketers with a more precise tool—not just for understanding how messages are received, but also for designing effective campaigns. What follows is a deep dive into this model: a sharper lens for evaluating and engineering messages that pass the brain's primal gatekeeper.

Primal Brain: The brain's first responder.

This is your ancient neural security guard. It is fast, automatic, and survival-driven. When it encounters a stimulus—like a looming

Figure 19: The First Responder Brain

deadline, a loud noise, or a price tag—it asks: Is this a threat? Can I eat it? Should I run? It processes sensory cues before you even realize what is happening. It is non-verbal, deeply emotional, and hates ambiguity. It loves contrast, simplicity, and safety. When your heart races after hearing a sudden honk, that is the primal brain initiating a cascade of arousal before your rational mind even names the sound as a car.

This primal system does not "think"—it reacts. As LeDoux's research revealed, emotional reactions can occur even without conscious recognition of a stimulus. Emotional hijacking is more than a metaphor; it is a neurobiological phenomenon. To fully grasp how persuasion unfolds in the mind, we must delve into the anatomy of this old brain—mapping out its regions in terms of their persuasive potential.

Core Primal Brain Areas

The emotional lift we seek in persuasion is orchestrated deep within the brain's oldest and most influential structures. These regions act more like sentinels—scanning for danger, novelty, or emotional intensity. Each plays a distinct yet interconnected role in how we evaluate, remember, and act on persuasive messages.

Often dubbed the brain's emotional radar, the *amygdala* plays a crucial role in detecting and amplifying relevance. The word *"amygdala"* comes from the Greek *"amygdalē,"* meaning "almond" (or amande in French, my native tongue). Early anatomists, upon dissecting the brain, noticed a small, almond-shaped cluster of nuclei deep within the temporal lobes. With no idea what it did at the time, they named it for its appearance. Today, we know the amygdala plays a significant role in processing emotional stimuli—especially fear, urgency, and other survival-relevant cues. So, in a poetic twist, the "almond" helps us decide what is emotionally nutritious enough to store or act on. Whether a message sparks fear, joy, or urgency, the amygdala sounds the alarm and influences both memory encoding and bodily readiness.

Meanwhile, the *insula* serves as a bridge between emotion and body awareness. It interprets internal cues—such as pain, temperature, hunger, and even cravings—and translates them into intuitive judgments. When messaging taps into bodily metaphors ("feel the burn," "taste the adventure"), it is the insula that makes those calls resonate.

But more structures play a critical role in the primal brain. For instance, the *brainstem* and *hypothalamus* regulate the machinery of arousal, including heart rate, pupil dilation, and hormonal release. These structures trigger immediate action tendencies, which is why countdown clocks and "act now" messages can quite literally raise your pulse. Also, the *hippocampus* acts as the brain's archivist, tagging emotionally significant stimuli and storing them in episodic memory. The word hippocampus originates from the Greek words "hippos" (meaning "horse") and "kampos" (meaning "sea monster"). Why? Because the structure's curved shape reminded early anatomists of a seahorse. Despite its mythological name, this brain region is essential for a very grounded function: memory encoding. It enables us to navigate time and space by stitching together emotionally significant episodes, creating the autobiographical narrative of our lives. If your message carries novelty or delivers a memorable surprise, it earns a place in this neural ledger—making future recall and brand association far more likely.

Central to reward anticipation is the nucleus accumbens (NAcc), often referred to as the brain's pleasure and motivation center. It responds to cues that signal potential rewards—whether social (praise), material (discounts), or experiential (excitement). When messaging creates a sense of "something good is coming," the NAcc helps drive approach behavior. It is why loyalty programs, gamified experiences, and limited-time offers feel irresistibly enticing: the dopamine spike in this area makes the future reward feel emotionally urgent.

Together, these regions embody persuasive messages. They ensure that what is emotionally relevant becomes physiologically urgent, behaviorally motivating, and unforgettable.

Arousal, Valence, and Salience

As noted already, the critical feature of the primal brain is its sensitivity to arousal and valence. *Arousal* refers to the intensity of emotional activation—whether a message triggers excitement, dread, or anything in between. *Valence* refers to the emotional direction, which is either positive (approach) or negative (avoidance). High-arousal emotions—such as fear, surprise, and joy—are far more likely to drive action. This is why emotionally charged ads consistently outperform rational ones (Venkatraman et al., 2015). The emotional lift a message creates—the delta between baseline emotion and post-exposure emotion—acts like a neural highlighter, guiding attention and enhancing memory formation. And *salience* is what gets encoded. The part of the brain responsible for filing away emotionally significant moments is the hippocampus, which we discussed earlier.

For example, a campaign that evokes warmth and surprise—like Coca-Cola's holiday ads—activates the hippocampus by creating emotionally charged novelty. The hippocampus responds not to logic, but to narrative and sensory immersion. It flags those moments that feel new yet emotionally familiar, encoding them deeply in our episodic memory. When the hippocampus tags a brand with such resonance, it turns a one-time impression into a lifelong mental bookmark.

Neuroscientist David Eagleman, in his book *Incognito* (2011), reinforces this by showing that primal (unconscious) processes make most decisions before our conscious mind ever catches up. He believes that our subjective sense of "deciding" is often just a story constructed after the neural die has already been cast. Eagleman explains that our unconscious mind processes millions of bits of data in parallel—something our conscious mind could never achieve. So, we may believe we are rational agents, but the truth is we are post-hoc rationalizers of emotional and instinctive choices. Our marketing messages, then, must appeal first to those shadowy, intuitive processes Eagleman so eloquently unpacked.

In persuasion, then, we are not just transmitting information. We arouse, contrast, and nudge action. And the salience of our messages lives or dies by how well we can engage our audience's hippocampus.

Rational Brain: The patient librarian.

While the primal and emotional systems light the spark of persuasion, the rational brain often intervenes with the fire extinguisher—or, more aptly, the narrative pen. We like to believe our choices are driven by careful reasoning, but neuroscience tells a different story. The rational brain and its connected regions help us justify, not initiate, most decisions. They integrate data, evaluate trade-offs, inhibit impulsive reactions, and weigh moral or social consequences. In short, they give our instincts a story that sounds logical enough to accept.

Yet reason is not irrelevant. It is just not in the driver's seat. When emotions have opened the gate, the rational brain refines the path forward. Understanding how these higher-order regions contribute to persuasive outcomes is crucial for crafting messages that are not only emotionally engaging but also cognitively satisfying.

Importantly, the entire rational brain is metabolically expensive.

Figure 20: The Patient Librarian

Engaging in reflective, logical thought consumes a substantial amount of glucose, attention, and cognitive resources. The brain, being the ultimate efficient machine, avoids invoking this slow system unless necessary. Therefore, unless a stimulus has successfully captured attention and passed the primal brain's filter, the rational brain will not even register it. In practical marketing terms: if the primal brain refuses to "greenlight" your message, your logical arguments, product specs, or unique value propositions never even get their day in court.

The rational brain is also the seat of narrative construction. It is where humans synthesize meaning. We link events, form stories, justify behavior, and create coherence. This is the domain of brand narrative and rational justification. For example, after emotionally desiring a luxury car (thanks to the primal brain), the rational brain chimes in with logic: "It is safer," "It holds value," or "I deserve this after a tough year." The decision was made emotionally; the rational brain merely filed the paperwork.

Meanwhile, to fully grasp how persuasion unfolds in the mind, we must delve more into the anatomy of the rational brain—mapping out its regions in terms of their persuasive potential.

Core Rational Brain Areas

First, let us discuss the critical importance of your *prefrontal cortex* (PFC), located in the frontal region of your brain and responsible for higher-order functions, including reasoning, planning, future projection, and impulse control. It is the part of the brain responsible for pondering, postulating, and parsing nuance. But it is slow and deliberate by design. While the primal brain is like an instinctive bouncer that lets or blocks entry in seconds, the PFC is more like a judicial panel, reviewing documents and holding deliberations. It does not act unless it absolutely must. Think of it as the control tower of your brain. It manages long-term planning, rational analysis, inhibition, and moral reasoning. It also manages what Antonio Damasio calls "conscious deliberation"—evaluating options not just for their logic but for their consequences. The PFC helps us justify that premium coffee maker you just purchased after the dopamine rush has passed.

Meanwhile, the *ventromedial prefrontal cortex* (VmPFC) is responsible for integrating the value (also known as the *utility* of your choice, an important concept we will unpack later) and personal relevance of a message. Situated in the lower middle portion of the prefrontal cortex, the VmPFC integrates emotional and cognitive inputs to assign subjective value. It is essential for answering the question: "What does this mean for me?" In persuasion, the VmPFC acts as a convergence zone, merging instinctive responses with contextual cues to produce a coherent judgment. It is the cortex's valuation engine, and it lights up when a choice feels personally resonant.

Another critical area, called the *orbitofrontal cortex* (OFC), mediates the assessment of social norms and rewards. Positioned just behind the eyes, the OFC evaluates reward considering social context and learned associations. It is sensitive to fairness, desirability, and reputation. In persuasive messaging, the OFC is particularly engaged by social proof (e.g., testimonials), status signals (e.g., luxury branding), and framing effects (e.g., "most popular choice"). It refines raw reward signals into socially calibrated preferences.

Nestled between the two hemispheres, another critical area called the *anterior cingulate cortex,* or ACC, detects when actions or choices conflict with goals, norms, or expectations. It plays a key role in resolving cognitive dissonance. In persuasive settings, the ACC may become active when a consumer encounters new information that contradicts existing beliefs—or when an offer appears to be "too good to be true." Effective messaging often anticipates and resolves these moments of friction, guiding the decision smoothly forward.

Finally, the *inferior frontal gyrus* (IFG) is primarily involved in language processing and narrative framing. Part of the broader Broca's area, crucial for speech production and language processing, the IFG is involved in language comprehension, metaphor processing, and syntactic structure. This region helps translate complex information into digestible chunks. Persuasive narratives, rhetorical devices, and well-structured taglines engage the IFG, allowing language to serve not just as a conveyor of facts but as a catalyst for cognitive fluency and emotional resonance. Additionally, the posterior parietal cortex (PPC) is responsible for allocating attention and performing spatial reasoning. Located near the upper rear of the cortex, the PPC supports attentional shifts and spatial orientation. It determines where cognitive resources are directed in a visual scene or interface. In marketing design, elements like layout, white space, and motion can engage the PPC, influencing what the primal brain has noticed first—and what gets remembered.

These cortical structures are the rational brain's instruments of interpretation. They do not override emotional systems, they translate them. In the symphony of persuasion, the cortex does not play the first note, but it helps write the story that follows. This brings us to a deeper layer of interpretation: the biases that shape what the brain pays attention to, what it remembers, and how it decides. If the cortex is the translator, then cognitive biases are its shorthand—neural scripts that fast-track meaning-making without exhaustive analysis.

Biases: The Brain's Shortcuts to "Yes"

We noted earlier in our discussion that Daniel Kahneman's work on dual-system thinking set the stage for understanding why many of our decisions are made without conscious deliberation. Biases are not mental glitches. Often referred to as "cognitive biases", I feel compelled here to remind the reader that most of them are deeply rooted in the role our primal brain plays in accelerating many of our decisions. That is why I considered all biases survival algorithms—evolved heuristics designed to speed up decision-making in an unpredictable world. Long before we had spreadsheets and split tests, our ancestors faced life-and-death choices where hesitation could mean extinction. Thus, evolution favored brains that made rapid, efficient judgments based on limited information. Curiously, biases persist in modern consumers, subtly steering attention, judgment, and purchase behavior—often beneath conscious awareness.

Thus, in the world of persuasion, biases are powerful levers. When marketers learn to align messages with these built-in tendencies ethically, they can unlock decisions that feel intuitive to the consumer but are neurologically orchestrated. Let us explore the most impactful biases at play during persuasive events—and how they connect to the primal and rational systems described in the NeuroMap® framework.

Anchoring Bias: First Impressions Anchor Value

Anchoring occurs when the brain fixates on the first piece of information it encounters, disproportionately influencing subsequent judgments. Display a $999 laptop next to a $1,499 model, and suddenly the cheaper one feels like a bargain—even if both are overpriced.

This bias is also rooted in the primal brain's reliance on reference points to assess survival value. When early humans encountered food in a scarce environment, knowing what "a lot" looked like mattered. Anchoring leverages this neural architecture by presenting "original" prices or extreme options, making the intended choice feel safer or more reasonable.

Social Proof: Safety in Numbers

The limbic system, particularly the amygdala, is finely attuned to social cues. Belonging equals survival in tribal brains. That is why we tend to conform to what others are doing, especially in ambiguous situations. The sight of 350 five-star reviews does not trigger rational analysis. It triggers a sense of belonging and reassurance: "This many people cannot be wrong."

In marketing, social proof functions as a neurological shortcut that reduces uncertainty. It suppresses cognitive dissonance and enhances emotional security, both of which are crucial for moving a consumer from hesitation to commitment.

Scarcity Effect: The Brain's Alarm on Loss Potential

A "limited-time offer" or "only 2 left in stock" activates the amygdala and other limbic structures that prioritize urgent action. This is about survival calculus. If a resource is scarce, the primal brain assumes it has value and demands immediate decision-making. Scarcity messaging also amplifies attention, a key factor in achieving persuasive impact.

Loss Aversion: Avoiding Pain Over Gaining Pleasure

Rooted in Kahneman and Tversky's Prospect Theory, loss aversion refers to the brain's stronger reaction to potential losses than to equivalent gains. Losing $100 feels twice as painful as gaining $100 feels pleasurable.

This bias plays directly into the NeuroMap® principle of identifying consumer "pain." Messages that highlight what consumers stand to lose—comfort, time, status, or money—if they do not act are neurologically more compelling than those that merely promise gain.

Status Quo Bias: The Brain's Default Mode

The primal brain favors the familiar because. The brain burns too many calories evaluating change, so unless there is a compelling trigger (such as pain, fear, or urgency), it prefers the default, which is to avoid it. This bias explains why even clearly better alternatives are often ignored. To counter the status quo bias, effective messaging must activate potent primal triggers—contrast, urgency, and relevance. Without these, your pitch will feel like background noise in a world already saturated with stimuli.

Confirmation Bias: Seeing What We Want to See

Confirmation bias causes us to seek out and recall information that aligns with our existing beliefs. From a neural standpoint, this maintains cognitive efficiency and emotional comfort. The neocortex collaborates with limbic memory centers to reinforce familiar patterns and reject threatening or incongruent data. This bias shapes how consumers interpret brand messages. Once they believe a product is "eco-friendly" or "luxurious," they will selectively process supporting evidence while downplaying contradictions. This is why consistency in brand positioning is neurologically essential.

Framing Effect: Context Changes the Choice

The framing of a message—its context, wording, and structure—can drastically alter perception. A yogurt labeled "90% fat-free" outsells one labeled "10% fat," even though both convey identical content. The limbic brain processes a negative frame faster since it is more urgent to avoid pain than to seek gain. However, positive framing appeals to the primal brain if it is emotional, concrete, and visually evocative.

Availability Heuristic: The Most Vivid Wins

Our judgments are skewed by what is most easily recalled, not by what is most accurate. Dramatic or emotionally salient examples dominate decision-making. This is why testimonials with vivid stories outperform dry statistics, and why a single image of a suffering animal can raise more funds than a report of mass extinction.

The primal brain is inherently visual and emotional. Make your messages easy to remember—and hard to ignore—and you will harness the full force of the availability heuristic.

The Bandwagon Effect: The Fear of Missing the Herd

Distinct from social proof, the bandwagon effect is about the *momentum* of popularity. "Everyone is switching to X" is more than a social cue—it is a signal that you might be left behind. Neurologically, this taps into both the primal brain's need for affiliation and threat detection: being excluded from the tribe once meant death. Marketers who frame trends as fast-moving trains that you cannot afford to miss are activating this ancient fear of being left out.

Temporal Discounting: Now Beats Later

When given the choice between a smaller immediate reward and a larger delayed one, most people tend to choose the former—irrationally but predictably. This is known as *temporal discounting* bias. The primal brain favors immediate gratification because, in evolutionary terms, the future is uncertain—and potentially dangerous. Waiting can mean missing out, or worse, not surviving. This bias explains why limited-time bonuses, instant delivery, or immediate access have such persuasive power. If your offer satisfies an urgent, immediate need, the primal brain listens. Deferred benefits, no matter how logical, struggle to compete.

This effect was famously illustrated by the *Stanford marshmallow experiment* (Mischel et al., 1972), where children were given the choice between one marshmallow now or two if they waited 15 minutes. Most opted for the immediate reward. Decades later, follow-up studies found correlations between those early choices and a range of life outcomes, including academic success and health indicators. These findings suggest that our ability—or inability—to delay gratification is deeply rooted in the brain's wiring and predictive models of the future. In marketing terms, the marshmallow test highlights a fundamental truth: when crafting persuasive offers, immediacy is a biological imperative.

The IKEA Effect: Effort Equals Value

It may seem paradoxical, but people tend to value things more when they have had a hand in creating them—even if that contribution is modest or inefficient. This phenomenon, known as the *IKEA Effect*, was named after the popular Swedish retailer whose flat-pack furniture requires customer assembly. In a 2011 study by Norton, Mochon, and Ariely, participants who built simple IKEA products, folded origami, or assembled Lego kits consistently valued their creations more than identical versions made by others. The implication? Labor leads to love.

From a neuroscientific perspective, this bias taps into the primal brain's reward circuitry. The act of investing effort—whether physical, cognitive, or emotional—activates dopaminergic pathways associated with ownership, pride, and satisfaction. The primal brain interprets that effort as a form of commitment: "If I worked on this, it must be worth something." This is an evolved heuristic; in ancestral environments, things we expended energy to acquire or create were crucial for survival and deserving of protection.

In marketing, this effect is gold. Brands that invite consumers into the creative process, whether through build-your-own tools, customization interfaces, or interactive content, increase perceived value. When users personalize a product, design their own packaging, or configure features to fit their identity, their emotional attachment deepens. The product becomes *theirs*. The IKEA Effect reminds us that persuasion is not always about making things easier. Sometimes, the brain assigns more value not in spite of the effort, but *because* of it.

To conclude, biases are the neurological architecture of persuasion. They allow the brain to navigate overwhelming choices with speed and minimal friction. When aligned ethically with consumer goals, these mental shortcuts can lead to more satisfying decisions, reduced regret, and greater brand loyalty. Biases are the brain's way of saying: "Help me decide fast and make me feel safe." Listen to that voice. Speak to it with empathy, clarity, and primal resonance.

Persuasive Implications: Speak the Brain's Language

Let us face it: the brain is not a computer, but rather a survival tool, and a meaning-maker. To persuade it, we must speak its dialect—emotion, urgency, clarity, reward. The message must first pass through primal filters of relevance, safety, and gain before rational logic even gets a chance to take the podium. Therefore, understanding the brain does not necessarily equate to manipulation—it is about seeking alignment. The more your message resonates with how the brain naturally operates, the more persuasive you become. Always remember that if your message fails to engage the primal brain, it will not reach consciousness. But if it respectfully and ethically aligns with our deepest fears, desires, and survival codes, you are connecting with it in milliseconds.

Neuroscience reframes the satisfaction or value derived from a choice through a biological lens: as briefly mentioned earlier, economists refer to it as the "utility" of a decision. Indeed, decisions are filtered through an evolved sense of cost and benefit, safety, and reward. Every time we make a choice—whether to click, to scroll, to buy—we are unconsciously calculating expected emotional reward against anticipated risk or discomfort.

This mental effortless calculus is deeply emotional and heavily influenced by context, memory, and experience. Antonio Damasio's somatic marker hypothesis demonstrates that we rely on emotional tags—stored in regions such as the ventromedial prefrontal cortex—to inform our decision-making. From a marketing perspective, this means that utility is about assessing the affective (note the spelling, affect is the study of our emotions) impact of a choice. Will this product make me feel safer, more competent, more attractive, more in control? Will it spare me embarrassment, give me status, or help me feel a sense of belonging? These questions are rarely spoken aloud but dominate the background processes of the decision-making part of the brain.

It is precisely this hidden choreography of choice that drew me to the work of Dan Ariely, Barry Schwartz, and John Medina early on. Their research not only confirmed what I was observing in neuromarketing, but it also gave it structure, vocabulary, and empirical grounding. That is why I find it essential to close this chapter with a nod to their pioneering insights.

The End of Rational Marketing

As I have argued for well over twenty years, I believe an overwhelming number of marketers have operated under a dangerous illusion: that humans are rational actors, capable of consciously articulating their preferences, comparing features, and making decisions based on logic. This belief is not just outdated—it is simply delusional. The reality behind our choices is far simpler and more fundamental.

Modern neuroscience has made one thing clear: most decisions are made emotionally, instantly, and unconsciously, then later rationalized to sound "reasonable." Yet marketers continue to rely on messaging that targets the rational brain—charts, bullet points, feature comparisons—as if persuasion were a courtroom trial. They continue to expect individuals to describe something they cannot see: the trustworthy source of their decisions, buried in the deep recesses of the primal brain. Also, despite overwhelming evidence from neuroscience and behavioral economics, many marketers and advertisers continue to rely on outdated, rational tools to decode consumer behavior: self-reports.

Indeed, surveys, focus groups, and sentiment analyses remain dominant—despite their dependence on conscious self-reporting, which captures only a fraction of actual decision-making processes. According to Greenbook's 2023 GRIT report, over 85% of marketers still use surveys as their primary insight tool, while 60% routinely rely on focus groups. These methods assume that what people say reflects what they believe and what they will do.

Meanwhile, popular tools like sentiment analysis and Net Promoter Scores (NPS) still rely on rational expression, providing limited access to emotional drivers that truly shape behavior. While some in the industry acknowledge these limitations, the adoption of biometric, neuroscientific, or implicit association techniques remains below 20% globally. In other words, most of today's marketing strategies are still built on data that the brain's storytelling centers fabricate after decisions are made—not before.

This is the great blind spot of rational marketing: it assumes that people are aware of their own motivations. They are not. Asking consumers why they chose something is like asking a fish to explain water. They will discuss price, quality, or convenience—but those are just post-hoc narratives, not the causes. Emotions, biases, instincts, and context drive behavior. Those forces operate beneath the surface of awareness.

In this last section, I continue to dismantle the myth of the rational consumer by drawing on the insights of three key voices in behavioral science and neuroscience: Dan Ariely, Barry Schwartz, and John Medina. Their work reveals just how misguided our faith in rationality has been—and why marketing strategies built on logic are not only ineffective, but they are also fundamentally misaligned with how the brain works. The time has come to stop selling to the brain we *wish* people had—and start speaking to the one they use. This is a biological fact.

Let us start with Dr. Barry Schwartz. In his groundbreaking book *The Paradox of Choice* (2004), Schwarz illuminated how an abundance of options—once considered a hallmark of consumer empowerment—can paralyze rather than liberate. His research shows that the more choices people face, the more anxiety, regret, and dissatisfaction they experience. In a world oversaturated with SKUs, product lines, and digital interfaces, cognitive overload activates stress circuits in the amygdala and anterior cingulate cortex, not the deliberative prefrontal cortex. The primal brain does not celebrate infinite freedom—it recoils from it.

Schwartz's research delivers a sharp rebuke to classical economics, which treats consumers as utility maximizers capable of comparing all options and making rational trade-offs. In practice, most people either choose impulsively or not at all. We all default to heuristics—gut-level shortcuts—precisely because our rational brain is easily fatigued and fundamentally overwhelmed by complexity. Schwartz demonstrated that fewer curated options, particularly those that tap into identity, emotion, and social belonging, yield better outcomes than encyclopedic menus of possibilities. His conclusion is radical (even more than mine): freedom is not maximized by more choices—more meaningful ones optimize it. Meaning is never computed—it is *felt*.

Then, meet Dr. John Medina (2008), whose *Brain Rules* distills neuroscience into 12 transformative insights, including the revelation that emotion is the gatekeeper of attention and visual stimuli are processed 60,000 times faster than text. These are not cute trivia points; they are strategic imperatives for modern persuasion. Medina demonstrates that we remember emotionally charged and visually engaging information far more than anything presented in dry, logical form. For example, his work shows that people retain only about 10% of information after three days if it is presented as text, but 65% if the duplicate content is paired with a vivid image.

Medina also underscores the evolutionary primacy of movement, threat detection, and pattern recognition—functions housed in the primal brain. He asserts that we are wired to survive, not to reason. For him, survival depends on quickly filtering stimuli for emotional salience. If information does not feel relevant, it does not get in.

Finally, Dan Ariely's research in Predictably Irrational (2008) exposed how reliably irrational our decision-making really is—and how those irrationalities can be used to steer behavior ethically. For instance, the *decoy effect* shows that people change their preference between two options when a third, inferior option is introduced. Ariely demonstrates that perceived value drives the shift, not logic. Meanwhile, his experiments confirm that losses loom larger than equivalent gains. According to his studies, a $100 loss feels twice as painful as a $100 gain feels pleasurable. This primal aversion is a core reason "do not miss out" messaging works: it frames inaction as loss.

Together, these three scholars present a unified argument: The primal brain is not just a passenger on the decision-making journey—it is the driver. Rational appeals may look impressive in a pitch deck, but they fail in the wild. Consumers are not evaluating features, specs, or benefits with Spock-like detachment—they are scanning for safety, status, novelty, and emotional resonance. Ariely exposed the irrationality. Schwartz revealed the emotional cost of too much logic. Medina anchored it all in brain science. The evidence is overwhelming and, frankly, irreversible: Rational marketing belongs to a dead era.

Before persuasion can dance with logic, it must knock at the door of survival. In this next chapter, we continue our journey into the primal brain relic of evolution that continues to influence decision-making. You will discover the twelve primal triggers—neurological shortcuts that seize attention and stir action. Be prepared to speak the language your audience's brain was wired to understand long before it could form a sentence.

Chapter Summary

This chapter introduces a vital shift in perspective for any persuasive communicator: stop thinking like a seller and start thinking like a brain. To influence behavior, one must first understand how the brain processes information, assigns value, and decides what to act on. This chapter goes beyond abstract neuroscience and provides a functional map—the primal and rational brain model—that helps marketers locate the authentic sources of influence inside the consumer mind.

The primal brain is reintroduced here as the brain's first responder. It is ancient, fast, and fiercely focused on survival. It filters all incoming stimuli through the lenses of fear, safety, pain, and pleasure. If your message does not immediately signal relevance to survival—whether physical, social, or emotional—it gets discarded before conscious awareness is even activated. This is why traditional rational messaging so often fails. The brain does not think its way into action; it feels its way toward safety and away from risk.

Emotion is the precondition for any persuasive effect. This is the territory of the limbic system, which works inside the Primal Brain to encode emotional intensity, trigger memories, and signal importance. A message that lacks emotional salience is invisible to the brain's decision systems. This explains why consumers often describe liking or trusting a brand without being able to articulate the specific reasons. Their emotional systems made the call, and their rational minds merely tagged along.

In contrast, the rational brain (the neocortex) is slower and more deliberate. It excels at abstract thinking, language, planning, and logic. However, it is also effortful and energy-intensive. As such, it only engages when the primal brain and emotional centers permit a message to pass through. This sequential pathway highlights why rational arguments are necessary but not sufficient. The brain's hierarchy favors instinct over intellect, especially when time or attention is limited.

To navigate this dualistic structure efficiently, the brain uses its "internal profit calculator"—a neural system designed to assess utility constantly: *Will this help me? Will it hurt me? Is the reward worth the risk?* These evaluations happen automatically and unconsciously. Understanding this helps marketers focus less on features and more on perceived gains: time saved, status earned, comfort gained, or pain avoided.

The chapter also introduces the metaphor of the "patient librarian" to describe the Rational Brain. It files, organizes, and explains—but rarely initiates on its own. Most decisions are already made before logic gets involved. In marketing terms, this means that post-rationalization is more than persuasion—it is a form of storytelling that occurs after the sale has already taken place in the subconscious.

To drive the final nail into the coffin of rational marketing, we turn to the work of three influential scholars: Dan Ariely, Barry Schwartz, and John Medina. Ariely reveals how irrational forces, such as loss aversion and the decoy effect, routinely hijack our choices. Schwartz demonstrates how too many options lead not to better decisions, but to paralysis and dissatisfaction. Medina shows that attention and memory are governed not by logic, but by emotional relevance and visual impact. Together, they expose the illusion of the rational consumer and the futility of persuasion strategies that rely on what people say, rather than what their brains do.

Despite this evidence, the marketing world remains heavily invested in self-report tools such as surveys, focus groups, and sentiment analysis, which depend on the conscious mind to explain decisions it did not make. Most marketing research, in short, still asks consumers to explain what their primal brain has already decided—without their awareness.

By the end of this chapter, one truth becomes inescapable: effective persuasion does not begin with logic. It begins with a sense of survival urgency, emotional resonance, and cognitive ease. Marketers must craft messages that pass through the primal gate morally and ethically, stir emotional currents, and only then appeal to the neocortex for justification. This brain-based blueprint serves as the foundation of NeuroMap®, and all subsequent persuasive strategies outlined in the book. Influence, in other words, is permitted by instinct.

A Quick Primal Exercise for You

The next time you analyze a marketing message—whether yours or someone else's—run it through this primal checklist: Does it grab attention? Look for contrast, novelty, or high-arousal emotional cues.

Is it emotionally engaging? Does it tap into joy, fear, desire, or urgency?

Is it simple and concrete? Can the primal brain process it without confusion or effort?

Does it promise immediate relevance or reward? Will it help the brain's host feel safer, happier, more connected, or more in control?

Would this message have mattered 100,000 years ago? If it appeals to survival instincts, it will resonate today.

Final Thought: Tapping into Fluency

Understanding the brain does not imply you are gaming the system; instead, it means you are leveraging the power of fluency. When you communicate with the brain on its terms, you *replace resistance with resonance*. Messages that follow the brain's rules do not shout—they harmonize. The art and science of persuasion, at its deepest level, is biological empathy. You are not just trying to get someone to say yes. You are inviting their brain to recognize it already has.

CHAPTER 4: The 12 Primal Triggers

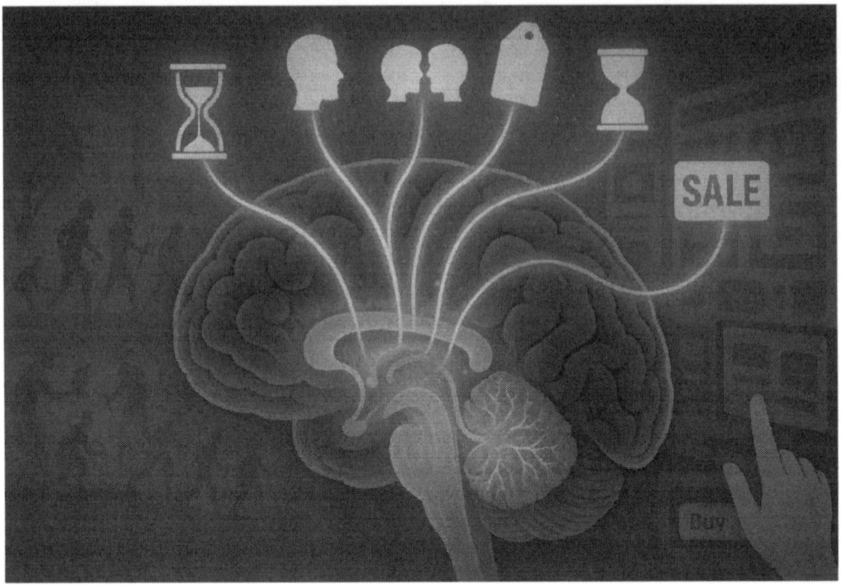

The wiring of the brain at the level of the amygdala allows emotional stimuli to be processed before they reach the cortex.

Joseph LeDoux, Ph.D.
Author of The Emotional Brain and Anxious

We live in an age of unprecedented information overload. The average person is bombarded with well over 30,000 marketing messages per day. To grasp what a day looks like inside our brain, imagine this: before your feet even hit the floor in the morning, your smartphone lights up with half a dozen app notifications—each one subtly branded. As you scroll through your newsfeed, every third post is a promoted ad, and every influencer is wearing, drinking, or endorsing something. On your cereal box, the packaging shouts "gluten-free" and "heart healthy," while your milk carton pitches sustainability. The radio or streaming service interjects ads between your morning playlists. By the time you commute, you have passed dozens of logos—on buses, billboards, bumper stickers, gas pumps, and coffee cups.

Throughout your workday, you continue to encounter email banners, LinkedIn ads, Google search results with sponsored placements, Slack integrations featuring brand mentions, and background visuals in Zoom calls that promote products. Even your desktop wallpaper might carry a subtle watermark from a design tool. The "splashing effect" of advertising is relentless.

By evening, TV commercials, product placements in shows, YouTube pre-rolls, TikTok hashtag challenges, and "recommended for you" pop-ups multiply. Even "free" apps bombard you with micro-ads tucked into corners, push notifications, or in-app purchases. In total, every branded jingle, notification badge, package label, tweet, sponsored meme, and influencer grin adds up. Not all are consciously noticed—but each one fights for a neuron, a glance, a micro-moment of attention. Each of these messages, whether a bold billboard or a subtle app notification, demands something from your brain. Few grab your full attention and enter conscious awareness, while 99% of the others are barely noticed, slipping through at the edge of perception.

But regardless of how aware you are, every single one draws on limited neural resources: oxygen and glucose. These messages compete for your brain's energy, and only the most relevant or emotionally charged stimuli make it past the primal filters. Most are discarded in milliseconds—but not before incurring a small toll in metabolic effort for your brain. Basically, your brain has evolved to act more like a border patrol agent than a sponge. It filters, scans, and ignores with ruthless efficiency. It lacks the time to ponder. Its job is survival, not appreciation. In less than a second, it must answer: Is this relevant? Is this dangerous? Is this useful? Anything that does not cut is discarded, often before the neocortex is even consulted. If persuasion is war, then attention is the battlefield.

So, in this chapter, we explore twelve powerful triggers that captivate the primal brain and bypass its filters. These are not gimmicks or tricks; instead, they are rooted in evolutionary psychology and cognitive neuroscience. Used ethically, they offer a blueprint for crafting messages that resonate, endure, and inspire action or decisions.

Trigger 1: Self-Relevance

Let us face it: the brain is a bit of a narcissist. It constantly asks, "What does this have to do with me?" Long before a message is analyzed logically, it is screened for personal relevance. Only when the answer is a resounding "everything" does the brain perk up and listen. Self-referential thinking, which plays a starring role in this process, has been shown to activate the medial prefrontal cortex (mPFC), a key region in the brain's decision-making and valuation systems (Northoff et al., 2006).

While not officially considered part of the ancient primal structures, such as the amygdala or brainstem, the mPFC still plays a unique bridging role. This part of your brain activates whenever something feels personally relevant or familiar. Moreover, what is remarkable is this: even though it is nestled in a more evolved region, the mPFC does not wait for a rational debate. It works in concert with primal areas to assign value and steer attention—especially when a message screams, "This is about *you*."

In practical terms, when the mPFC lights up, it signals a biological green light: the brain is now ready to encode, evaluate, and act. That is why self-relevance is more than a buzzword—it is a neurological ignition switch. Here are other key primal areas that show activation under self-relevant conditions:

- **The Amygdala**
  When a self-relevant stimulus carries emotional weight, particularly fear, threat, or reward, the amygdala becomes activated. This suggests that the brain treats personally relevant

stimuli as higher-priority *threats or opportunities*. It tags them for immediate attention and potential memory encoding.

- **Insular Cortex (particularly the anterior insula)**
  Although slightly more evolved than brainstem structures, the insula integrates internal bodily states—such as heartbeat, breath, and gut feelings—with emotional awareness. It lights up when something "feels" personal, especially in emotionally charged self-referential situations. It is a visceral alert system.
- **Periaqueductal Gray (PAG)**
  Part of the brainstem and essential for defensive behaviors, the PAG has been shown to activate in response to personally threatening or highly self-relevant survival cues (e.g., pain anticipation or social threat).
- **Hypothalamus**
  When self-relevant content intersects with needs like hunger, safety, or arousal, this ancient regulator can activate to modulate hormone release and autonomic responses.

In short, while the mPFC interprets the personal meaning of a message, other primal structures ignite the body's readiness to react. They prepare the organism—emotionally, physiologically, and motorically—for action. So yes, self-relevance does not only echo in the prefrontal cortex; it vibrates down into the body's ancient command centers, ensuring that what feels personal becomes physiologically urgent.

The bottom line is that the brain is a self-centered organ. From a survival standpoint, this makes perfect sense. If something is not about you, it is not worth your neurons. That is why messages that include personal language ("you," "your") or relevance to one's immediate goals are far more likely to gain traction.

In fact, Patrick and I wrote an entire Chapter on the power of "you phrasing" in the Persuasion Code. In it, we argue that the use of second-person language ("you") is more than a stylistic choice; it is also a neurological phenomenon. When messages use "you," they immediately become self-relevant, triggering activation in the medial prefrontal cortex (mPFC), our primary hub for self-referential processing.

Basically, persuasive messages must pass through an ego filter before any deeper processing can occur. Using "you" helps accomplish that. This is why the first step of our persuasion process is to: *Diagnose the Pain*. You cannot diagnose or re-awaken pain without speaking directly to the person experiencing it. Our model asserts that "you"-centered phrasing sharply contrasts with common marketing messages that use jargon, third-person framing, or abstract benefits, all failing to trigger self-interest or urgency. Instead of saying, "Our solution optimizes operations," we urge communicators to say, "You will stop wasting hours on manual work."

In a way, "you phrasing" is a linguistic mirror—it forces the audience to see themselves in the message. That is what gets through the primal gatekeeper. Without it, your message is just another in the endless queue of ignored noise.

Trust me, we all have an unconscious radar for content that matters to us. We tune out generalities and tune in to specifics that match our inner narratives. The more precisely a message aligns with our goals or anxieties, the more likely it is to bypass our primal brain's filters. Consider how medical advertisements personalize risk: "If you are over 50 and have a family history of heart disease…" That is primal targeting. It invites the brain to whisper: "This could be about me."

In practical terms, this means using audience research not only for segmentation but also for gaining empathetic insights. Speak their language. Reflect their reality. When people see themselves in your message, they are more likely to see the value in your offer. When you open with what your audience cares about—their pains, fears, and dreams—you can align with the primal drive to survive and thrive.

Trigger 2: Salience

We introduced this critical trigger in Chapter 3. Salience is the brain's mechanism for determining what is worth paying attention to. It refers to the degree to which a stimulus—whether emotional, sensory, or cognitive—stands out and commands a response from the brain, particularly the primal systems wired for survival. The more salient a message, the more likely it is to bypass the primal brain's filters and trigger immediate engagement.

This is not just about emotions, although they are a powerful form of salience. Emotional stimuli do activate the amygdala and limbic system, boosting the likelihood that a message will be remembered and acted upon (Phelps, 2004). Fear, joy, surprise, anger, and awe—all create spikes of attention and form durable neural impressions. But salience can be conveyed via vivid visuals, unexpected sounds, or ideas that challenge assumptions and create cognitive dissonance.

Neurologically, the salience network, anchored in areas such as the anterior insula and dorsal anterior cingulate cortex, acts as a switching mechanism between the brain's default mode network (DMN) and executive control networks. It is constantly on the lookout for novelty, relevance, or anything that threatens or excites. That is why a sudden color change, a surprising metaphor, or an emotionally provocative image can capture attention even in a sea of competing stimuli.

So, for marketers, salience means using a mental highlighter. It requires aligning your message with what the brain considers worth paying attention to. Is your message emotionally compelling? Is it sensory-rich? Does it introduce a conflict or promise a resolution? If not, it may be cognitively invisible—processed and promptly forgotten.

So, whether you are crafting a campaign, a product launch, or a TED talk, aim for salience, because the primal brain does not prioritize what is polite, expected, or technically perfect. It prioritizes what stands out and stays lodged in memory.

To deepen the discussion of salience, it is helpful to draw on Dr. Schwartz's teachings once again. While modern marketing often celebrates personalization and infinite selection, the primal brain rebels against complexity and the prospect of facing too many choices. It seeks clarity, not abundance. It craves binary cues: safe or dangerous, yes or no, approach or avoid. This is where salience becomes a neurological lifesaver—it cuts through noise by presenting stark contrasts that feel urgent, personal, and easy to act upon.

Consider this: when you are at the grocery store, faced with 15 types of olive oil, your rational brain may scan labels, prices, and origins. But what about your primal brain? It screams for simplicity. A label that says “Best for Heart Health” or “Chosen by Chefs” leaps out—not because it is more factual, but because it offers a clear, emotional, and biologically relevant shortcut. That is salience doing triage on your attention.

During my time as CMO of a grocery retail chain a couple of decades ago, I had a front-row seat to one of the most silent but costly phenomena in consumer behavior: cart abandonment in the aisles, not just online. I watched it unfold in real time—shoppers standing motionless in front of an endless wall of choices: 19 types of pasta, 34 variations of yogurt, a dozen identical brands of bottled water. Their hands would hover, eyes darting back and forth, shoulders tense. And then… paralysis. Some would eventually grab a default choice and move on. But many—too many—walked away. Left the aisle. Abandoned their carts.

They were not only tired. They were neurologically overwhelmed.

This was not a failure of product quality or shelf appeal. It was a failure of salience. The primal brain, confronted with too many similar stimuli and no clear point of contrast, disengages. It shuts down. Decision fatigue sets in, and with it, avoidance behavior.

In those moments, I realized that the choice abundance is a cognitive tax that no one is willing to pay. Real power lies in helping the brain make a fast, confident decision. That means distilling messages down to binary cues: *this helps you or it does not*, *this solves your pain or it adds to it*, *this is for you, or it is not*. Salience is about what breaks the tie in a brain begging for clarity.

When we started applying primal triggers like "best seller," "doctor-recommended," "quickest dinner solution," we saw movement. Not just product movement, but psychological momentum. Shoppers began to act, not freeze. Because when everything screams for attention, the brain listens only to what feels relevant, clear, and urgent.

In digital terms, consider a website pop-up: "Subscribe to our newsletter" is a cognitively dull message. But "Do not miss out—get the secret before it disappears"? That creates a binary emotional fork: *gain it now or lose it forever*. This taps into loss aversion, urgency, and social proof, all of which light up survival circuitry. That is what the brain notices.

Neurologically, the salience network activates in milliseconds when something violates expectations or sharpens focus through contrast. This is why a message framed as "Pain vs. Gain" or "Right vs. Wrong" often outperforms nuanced arguments. The primal brain does not like shades of gray—it wants rapid classification to preserve energy and safety.

So, when crafting persuasive content, do not just ask: *Is this true?* Ask: *Is this binary?* Is there a clear, emotional decision point? Will the brain label this as *a must-act or can-ignore option*? If not, your message may be accurate, but it will not be *salient*—and thus, it will not grab attention.

Trigger 3: Change

While salience refers to the degree to which something captures our attention, change is obviously one of the most effective drivers of salience. Though intimately related, these concepts are not interchangeable. Salience shines the brain's spotlight on what it deems important, whereas change acts as a crucial trigger to direct that spotlight in the first place. A message might be emotionally powerful, visually arresting, or intensely relevant (high salience) without necessarily introducing a significant shift from the brain's baseline expectations (change).

Biologically, the human brain is finely tuned to detect change. This primal skill developed under the evolutionary pressure of survival: early humans who could quickly notice subtle shifts in their surroundings—such as rustling leaves indicating the presence of a predator—had a far greater likelihood of survival. Anthropologist and evolutionary psychologist Robin Dunbar has extensively explored this adaptation, noting how heightened change detection became essential to human group dynamics, resource management, and threat response (Dunbar, 1998). Thus, our brains developed exquisite sensitivity to unexpected stimuli—loud noises, sudden movements, unfamiliar odors—all cues signaling potential danger or opportunity.

However, paradoxically, change is also the neurobiological enemy of the primal brain. Despite the evolutionary importance of detecting change, our ancient neural circuits inherently resist the uncertainty and risk that change often represents. Humans are hardwired with a deep, subconscious preference for predictability and stability, qualities that signal safety and familiarity. Neuroscientists, including Joseph LeDoux, highlight how our amygdala and limbic systems instinctively respond with stress, anxiety, or even fear when confronted with excessive novelty or unfamiliar scenarios, as these might represent potential threats (LeDoux, 1996).

This resistance manifests clearly in modern consumer behavior. Consider how difficult it is to break established purchasing habits or brand loyalties. Customers often instinctively shy away from new products, especially those that dramatically deviate from what is familiar or predictable. The success of long-standing brands, such as Coca-Cola and Levi's, exemplifies our primal craving for predictability; consumers prefer what they know. When Coca-Cola attempted its radical formula alteration in 1985 ("New Coke"), consumer backlash was swift and fierce, precisely because the primal brain interpreted this unexpected change as a threat rather than a reward (Gladwell, 2005).

From a neuroscientific perspective, as we have discussed earlier, the primal brain actively attempts to conserve cognitive energy; therefore, it favors known patterns over uncertainty. Daniel Kahneman refers to this as cognitive ease—our brains prefer familiarity because it is less metabolically taxing (Kahneman, 2011). Simply put, routine and stability preserve brain resources, while unpredictable changes consume energy and generate stress.

In stark contrast, the rational brain—the neocortex—actively seeks novelty, complexity, and change. This brain area thrives on new information because it fuels learning, growth, and innovation. The rational brain relishes prediction errors—the gap between expectation and reality—as primal dopamine-rich areas, such as the substantia nigra and ventral tegmental area (SN/VTA), fuel our appetite for learning and rewards (Lisman & Grace, 2005). Indeed, studies using *f*MRI scans show clear cortical activation in response to novel stimuli, demonstrating that our rational faculties crave and naturally seek innovation and variety (Berns, 2005).

This constant tension between our primal aversion to risk and our rational attraction to novelty shapes much of human behavior. Lifestyle habits exemplify this tension vividly: People tend to stay in unhappy relationships or unsatisfying jobs because, neurologically, the primal brain deems even uncomfortable familiarity safer than uncertain novelty. Yet, rationally, we admire those who bravely seek new careers, travel to unknown destinations, or innovate disruptive technologies precisely because the rational cortex is wired to explore and embrace the unknown.

What does this neurobiological dance mean for messaging and marketing? When something is too predictable or familiar, it no longer stimulates the brain's attention systems, particularly the primal brain that scans for risk. This means repetitive or overly familiar messages become invisible. However, it also means that excessive emphasis on novelty may spark anxiety: When a message, product, or experience is too unfamiliar or strange, the primal brain may interpret it as a potential threat.

To engage both the primal and rational aspects, you need to perform a *neurobiological balancing act*. Effective marketing must walk a tightrope: introduce just enough newness to spark interest (engage the limbic system and the primal brain) but anchor it in enough familiarity to avoid anxiety (so the neocortex does not reject it). That is why companies like Apple release innovations in *incremental* steps—new enough to excite, but familiar enough to feel safe. Effective advertisers (few and far between) intuitively understand this balance, which is why incremental updates, such as Apple's slight but meaningful product innovations, resonate far more successfully than radically unfamiliar designs.

In summary, while salience leverages the primal brain's instinctive alertness, change must be employed carefully, respecting primal resistance and rational curiosity. Overwhelming unpredictability triggers primal anxiety, while absolute predictability ensures boredom and disengagement. Thus, marketers, educators, leaders, and communicators face the constant challenge—and opportunity—of carefully managing novelty, understanding that while change is biologically uncomfortable, it is simultaneously irresistible to our rational, curiosity-seeking brain.

When crafting your next campaign, carefully calibrate this neurological equation: remain sufficiently familiar to maintain comfort when addressing the primal brain and deliver just enough change to make it compelling for its rational colleague. If your communication does not create this delicate ripple, it risks becoming ignored—just more noise in an overcrowded brain.

Trigger 4: Visual Priority

Our ability to see predates our capacity to think. Long before humans developed the sophisticated computational systems housed in the back of our neocortex, we were equipped with rapid and efficient primal visual processing tools that allowed us to detect motion, assess threats, and navigate our environments safely. This primal system is more than evolutionary baggage; it remains powerfully active today. Indeed, we are wired primarily for sight. Over half of the human cortex is dedicated exclusively to visual processing (Potter et al., 2014). In evolutionary terms, this immense cortical investment speaks volumes about the survival value of visual processing.

Consider early human survival scenarios. Individuals who spotted predators or rivals faster were those most likely to survive and pass on their genetic legacy. This evolutionary pressure has forged an enduring neural architecture that prioritizes visual information over other sensory inputs. Today, the legacy of this ancient urgency is visible in how swiftly and effortlessly our brains interpret visual data. According to MIT neuroscientist Mary Potter and her colleagues (Potter et al., 2014), the human brain can identify and classify images seen for as little as 13 milliseconds—a speed that far outpaces conscious thought.

Further reinforcing the dominance of visuals is the finding by neuroscientists that visual stimuli are processed 60,000 times faster than text. This astonishing speed advantage emerges because visuals bypass much of the brain's complex decoding processes required for language. Textual information must first be recognized as symbols, decoded into phonological sounds, matched with meaning from lexical memory, and finally integrated contextually. In contrast, visual processing often involves near-instantaneous recognition, particularly for images involving faces, emotional expressions, or high-contrast scenarios.

Numerous cognitive and neuromarketing studies (including mine!) back the potency of visual imagery. In groundbreaking research by neuroscientists such as Antonio Damasio, it was demonstrated that emotional visual imagery activates the amygdala and hippocampus—regions deeply involved in emotion and memory—far more efficiently than text (Damasio, 1994). Similarly, an influential study published in the Journal of Neuroscience found that images that elicit strong emotional reactions significantly enhance memory retention, compared to neutral or text-based information (Kensinger & Corkin, 2004).

Consider the iconic photograph "Tank Man" from Tiananmen Square, or the unforgettable image of the Syrian boy, Alan Kurdi, lying on a beach. These visuals captured global attention in ways that even lengthy, descriptive articles could not. Each image became instantly iconic and emotionally charged, deeply ingrained in the collective memory. The neural explanation behind this phenomenon lies in the primal brain's capacity to rapidly scan visual input, quickly identifying elements with emotional salience, threat potential, or significant contrast.

This insight is crucial for marketers: visual content is so much ore than decorative; it is essential neurological currency. The primal brain does not "read" text; it swiftly scans visuals to assess relevance and urgency. Faces, gestures, colors, contour, lines, and shapes all serve as cognitive shortcuts, instantly conveying meaning and evoking emotional responses. A single, striking image can communicate more immediate understanding and emotional resonance than an entire page of carefully crafted copy.

Companies like Nike masterfully exploit visual triggers. The simple image of a determined athlete crossing the finish line or an emotionally charged close-up of an athlete's face in action evokes immediate inspiration and aspiration, often far more persuasively than detailed textual descriptions of product features.

Moreover, modern digital marketers confirm these findings empirically. HubSpot research reveals that posts with compelling visuals receive 94% more engagement than those without. Likewise, BuzzSumo analysis shows that articles featuring an image every 75-100 words receive double the shares of those with fewer images. These results are not coincidental; they reflect how our brains are inherently optimized to engage more effortlessly and deeply with visual content.

Nevertheless, a cautionary note is necessary. Many marketers mistakenly equate aesthetically pleasing visuals with cognitive effectiveness. This misunderstanding overlooks the cognitive differences between reading and viewing. Reading text requires extensive neural resources, including phonological decoding, lexical retrieval, contextual integration, and sustained engagement of working memory. In contrast, visual perception leverages neural efficiencies honed by millennia of evolutionary necessity. Poorly designed visuals, cluttered layouts, or overly complex infographics can create cognitive overload, detracting from the inherent advantages that visuals offer.

Meanwhile, visual understanding further enhances persuasive power by capitalizing on our innate appetite for visual narratives that vividly highlight contrasts related to value: essential changes you promise to radically improve in a situation, such as eliminate pains versus deliver gains, the struggle of "before" versus the "joy" of after, good versus bad, or the weakness of your competition versus the strengths of your solution. By now, you recognize that humans are hardwired to seek out and rapidly understand such contrasts because they provide immediate cognitive clarity and emotional resonance.

For instance, mattress company Sealy famously released a split-screen advertisement showing, on the left, a young boy incessantly jumping on a poor-quality mattress (an annoyance). A few seconds later, to the right of the screen, the same boy only bounces once and abruptly falls asleep – presumably because the Sealy mattress is so effective at inducing and nurturing sleep. I have shown this ad to thousands of people at my conferences and workshops, which have been held in over 30 countries. Not only does it automatically generate explosive laughter, but it also instantly conveys Sealy's value proposition through clear contrast, effectively illustrating the difference their mattress delivers without requiring a single word of explanation.

Another memorable example is Weight Watchers' iconic advertising campaign, which humorously employed a split visual of a large door on the left side featuring a sign that says "entrance", juxtaposed with a smaller door on the right (and the word "exit" on it), visually dramatizing the brand's core promise—weight loss transformation. While compelling and memorable, it is important to acknowledge that by today's standards, such visual portrayals could be considered insensitive or politically incorrect, given the broader societal emphasis on body positivity and inclusivity.

Meanwhile, many advertisers and marketers tend to resist or avoid these techniques, mistakenly believing they are too simplistic or lack sophistication and creativity. This avoidance is a huge mistake! Indeed, the most profitable template in commercials, by far, is the narrative arc developed by infomercials—those simple, yet compelling 'before and after' stories or 'pain versus gain' scenarios. Despite people claiming in surveys and focus groups that they find infomercials annoying or simplistic, the truth is that these commercials consistently outperform the glitzy, expensive, over-produced ads frequently seen during high-profile events like the Super Bowl.

Many empirical and academic studies support this claim. Agee and Martin (2001) found that infomercials generate purchases that often involve planning rather than impulse, indicating deliberate consumer responses rather than spontaneous buys, which achieves the bottom-up effect of persuasion discussed in Chapter 2. Their follow-up research with Bhimy (Martin et al., 2002) demonstrated that TV spots featuring expert testimonials, demonstrations, and customer stories yield significantly higher response rates than traditional branding ads.

Direct-response television (DRTV) and infomercials also deliver measurable ROI advantages over most brand-centric ads. These formats allow advertisers to "buy responsiveness" instead of merely eyeballs (or the overused claim of raising awareness!)—translating audience engagement directly into sales or leads. For instance, a striking Kellogg Insight study analyzing over 200 consumer packaged goods campaigns—many of which attempted to sell cinematic narratives as if they were auditioning for an Oscar—found that the average elasticity was so low that doubling the TV ad spend produced only a 1% increase in sales. This sobering statistic highlights the underperformance of traditional brand ads in terms of ROI when compared to more direct, contrast-rich formats.

These findings clearly confirm the remarkable effectiveness of contrasted visual narratives in leveraging primal brain responses. This rapid comprehension bypasses neural friction and channels persuasive focus directly on value. Rather than relying on decorative and often distracting creative loops, these templates use visual dominance, saliency, and narrative structure to engage primal attention systems.

To sum up, this is why marketers should lean into this simple but powerful format:

1. **Visual explanations beat branding gloss**: Infomercials may lack glitzy production, but they drive measurable action—calls, clicks, conversions unlike many spots that focus on raising awareness.
2. **Neurological alignment**: Visual contrast and narrative arcs based on PAIN-CLAIM-GAIN align with innate brain mechanisms for meaning-making.

3. **ROI superiority**: Case studies show direct response formats converting at significantly higher rates than general brand ads, especially in ROI-measured campaigns
4. **Resilience over time**: Unlike time-limited branding ads, infomercials can continue to generate leads over extended periods—especially when cloned digitally or turned into evergreen content.

While marketers may rate infomercials low in focus groups, actual behavior tells a different story: infomercials are effective in selling products. And sell well. Despite claims of annoyance or inelegance, when executed with clarity, contrast, narrative structure, and emotional resonance, these formats outperform high-budget, fleeting brand messages.

Since its inception in 2002, my neuromarketing consultancy (SalesBrain) has consistently leveraged these visual principles, dominance, contrast, and saliency, to create visually driven templates that deliver rapid and impactful understanding. By designing visual frameworks that articulate clear contrasts, such as pain versus gain, before versus after, and competition versus proprietary solutions, SalesBrain has effectively enhanced or amplified the persuasive messaging for thousands of clients across over 600 industries.

One such campaign, produced for the Humane Society of the United States, underscores this principle perfectly. A revamped DRTV spot—tested against newer, more creatively ambitious versions—showed the limits of too much creative spin. We were tasked with neurotesting an old ad featuring stark, emotionally potent images of animal cruelty set to Sarah McLachlan's melancholic song, adhering to the classic pain-relief structure: sorrow followed by hope. Predictably, it outperformed a new spot that attempted bolder, more abstract promises, such as ending human mistreatment of all animals altogether. The lesson was both simple and profound: when you try to save poetic abstraction, you often lose the brain's attention; but when you show one wounded puppy and the potential to help it, you win the heart—and the donation.

If attention is the premium currency in our information-rich era, visuals are the bold font, capital letters that immediately draw the eye. From infographics distilling complex ideas into accessible formats to emotionally charged advertisements and storytelling through cinematic visuals, the potential to leverage this neural preference is profound.

Marketers seeking maximum persuasive impact must treat visuals not as supplemental, but as fundamental. The human brain's unparalleled efficiency in visual processing makes it clear: a picture is not merely worth a thousand words; it is worth a thousand neural connections firing rapidly and unforgettably. Visuals offer a direct line to the ancient pathways that dictate attention, emotion, and memory itself.

Trigger 5: First and Lasting Impressions

The brain is more likely to recall information from the beginning and end of a sequence (Murdock, 1962; Glanzer & Cunitz, 1966; Murre & Dros, 2015). Known as the "serial position effect," this phenomenon highlights a deeply rooted survival mechanism: start strong and end stronger. Our brains encode these moments more deeply than the middle, which often dissolves into cognitive fog. The hippocampus plays a starring role here, acting as the librarian of memory, paying special attention to what comes first and last.

However, this is not just about memory, but also about narrative structure. We are, at our core, storytelling creatures. Evolutionary psychologists argue that our brains evolved to favor narrative because stories were essential to survival. They conveyed danger, strategy, and social cohesion long before written language existed. Moreover, what makes stories memorable? Their beginnings and endings.

Neuroscientific research confirms that narratives engage a wider network of brain regions than raw facts alone (Green & Brock, 2000). When a story begins, our brains prepare to make sense of it. When it ends, they seek resolution and closure. This neurological bookending enhances retention and emotional impact. Stories are more than entertainment. They are evolution's original persuasion engine. Long before we had data points and dashboards, we had myths, parables, and cautionary tales—narratives that shaped behavior, cemented social norms, and ensured survival. Today, neuroscience confirms what the ancients knew intuitively: stories do not just inform; instead, they transform.

When a narrative is well-told, it triggers a neural symphony. The brain does so much more than processing the plot; it simulates the experience. Regions associated with empathy, motor planning, and even sensory perception light up as if the listener were living the story. This neural coupling—the synchronization between storyteller and listener—makes stories uniquely powerful in rewiring beliefs and motivating behavior.

At the heart of this transformation lies the limbic system, especially the hippocampus and amygdala, which encode emotional resonance and memory. When a story evokes emotion, it bypasses the analytical filter of the neocortex and secures a direct line to the primal brain. This is the exact mechanism leveraged in effective neuromarketing. Kahneman would say that it engages System 1 before System 2. Emotion first, logic second. Always.

But there is an additional—and often underestimated—mechanism at play: *narrative transportation* (Green, & Brock ,2000). Coined in communication science, this concept describes the immersive process by which an individual is "transported" into a story world. As transportation increases, resistance to persuasion decreases. The critical, skeptical mind loosens its grip, and the emotional mind takes the wheel. In this suspended state, the story becomes more than entertainment—it becomes a surrogate experience. Beliefs change, behaviors shift, and empathy expands. Studies have shown that the more transported someone is by a narrative, the more likely they are to adopt attitudes consistent with the story's themes.

This neurological and psychological dynamic is no accident. Narrative transportation engages both cognitive and affective systems, integrating attention, memory, emotion, and identity into a seamless process of meaning-making. It is, quite literally, a mechanism of neuroplasticity. Well-crafted stories forge new connections, overwrite outdated mental models, and offer the brain a new blueprint for interpreting reality.

In my latest book, *OPEN: A Neurospiritual Exploration of the Self-Healing Power of Your Brain*, I delve into this phenomenon more deeply—not just in the context of marketing or persuasion, but also in how narratives can facilitate psychological and physiological healing. Drawing on both neuroscience and ancient healing traditions, I demonstrate how personal storytelling, especially when grounded in vulnerability and coherence, can activate the brain's self-regulation networks and reconfigure entrenched neural patterns. Stories, it turns out, are not just persuasive; they are medicinal.

While our primary focus here is consumer behavior, the implications of storytelling are far beyond commerce. They touch identity, mental health, and even spiritual awakening. They remind us that the true power of persuasion lies not in control, but in connection, not in manipulation, but in meaning. In a world saturated with information, it is the emotionally resonant story—not the data point—that endures, persuades, and heals.

For marketers, communicators, and anyone trying to influence behavior, the implications are clear: treat your openings and closings like sacred ground. Do not start with fluff. Begin with tension, surprise, or self-relevance. End with clarity, emotion, or a call to action.

To truly persuade, it is clear again that we must first speak the language of the primal brain. This is not a metaphor. It is neurobiology. Remember that the primal brain is never swayed by data or logic. It is tuned to detect threat, novelty, and emotional salience. As David Ogilvy put it,

"When you sell fire extinguishers, open with the fire."

Often hailed as the "father of advertising, Ogilvy was a British ad executive whose timeless principles shaped modern persuasive communication. Founder of the legendary advertising agency *Ogilvy & Mather*, he championed emotionally resonant, research-driven advertising and famously insisted that effective marketing must first capture attention by speaking directly to human desires and fears. In other words, before you present a solution, ignite the problem. Wake the brain up. Feel the heat under the chair.

This is the first act of any brain-centric narrative: begin by re-awakening verifiable frustrations, blocks, pains, and needs that people find intolerable. Whether it is the fear of missing out, being excluded, failure, or staying stuck, pain captures attention—and it does so viscerally. For example, a fitness app should not begin with promises of six-pack abs; instead, it should start with the quiet shame of skipping workouts and the creeping anxiety of deteriorating health. Likewise, a cybersecurity firm should not sell software by boasting encryption specs; instead, it should dramatize the horror of waking up to a breached bank account or leaked customer data. Even a simple productivity tool like a calendar app can open with the nagging feeling of falling behind, missing meetings, or never quite gaining control of one's day. These scenarios all speak to primal fears—of being unsafe, inadequate, or left behind. When positioned as the opening emotional note, they activate the brain's survival circuitry. They signal to pay attention. This might matter more than you think.

Moreover, once that fire is lit, the story begins. Arousing attention—particularly through emotionally charged stimuli, such as fear—can indeed serve as a powerful entry point into the consumer's mind. However, this approach must be accompanied by a clear ethical boundary. The goal is to resonate, not to manipulate. Fear can be a door-opener, but it should never be weaponized. Marketers must avoid, at all costs, the trap of fearmongering and exaggeration, a regrettably common tactic among those who confuse urgency with alarmism.

I believe that a sense of conscience should accompany persuasion rooted in neuroscience. When we understand how deeply primal emotions steer decision-making, we also inherit a responsibility: to use that knowledge with empathy and restraint. Ethical persuasion is the highest form of influence. It respects the audience's cognitive vulnerability while guiding them toward decisions that serve both their needs and aspirations. Neuroscience provides a valuable perspective that should be applied with clear purpose, accuracy, and adherence to strict ethical standards. Through narrative transportation, stories invite the audience to step into a narrative world, experiencing emotions, conflicts, and resolutions as if they were reliving the experience in the present moment —the only time that matters for the primal brain. Neuroimaging confirms this: as people listen to an emotionally engaging story, their sensory, motor, and affective regions light up in synchrony with the storyteller's brain. The more they are transported, the less they counterargue. The story becomes a Trojan horse for belief change.

Within this narrative frame, introduce up to but **no more than three persuasive claims**. Why three? Because that is what the brain can comfortably hold and emotionally process under a light cognitive load. Recent experimental studies provide precise evidence that three items are often the effective ceiling for working memory under natural conditions. Typically, our brains use a trick called "chunking" to boost memory, for example, remembering the sequence 1-9-7-3 as the year "1973" rather than four separate digits. Tarnow (2016), in a study using *unchunkable* double-integer items, found that participants could reliably recall about 2.54 items when presented with three. However, performance deteriorated significantly when even one more item was added. Unchunkable means that the information received by our brain cannot be grouped into meaningful patterns, making it more challenging to recall.

So, when information is deliberately structured so that it cannot be grouped—such as two random double-digit numbers, say "38" and "71"—it makes the task of remembering each piece much harder, revealing the proper limits of our raw working memory. This suggests a hard capacity threshold around three distinct, ungrouped units, as recommended by NeuroMap®. Similarly, Cowan (2001) reconceptualized earlier models, such as Miller's "seven ± two," proposing that while short-term memory can sometimes manage up to four chunks, the focus of attention—the part of working memory used during real-time cognitive tasks—is limited to approximately three. Under conditions of cognitive demand, this number becomes the practical upper limit. Structuring persuasive messages around three core claims, then, reflects the brain's actual processing bandwidth. Beyond three, coherence fragments and cognitive load compromise retention.

Meanwhile, each one of your claims or taking points should address a distinct driver of motivation. The first might soothe the specific pain previously introduced. The second can build emotional resonance by connecting with a deeper identity or aspiration. The third should aim forward—toward relief, reward, or a vision of transformation. When presented within the story's arc, these claims do not feel like arguments. They feel like revelations.

Finally, the narrative must resolve not in abstraction, but in emotion. It should deliver what I call an *emotional lift*—a final surge of feeling that leaves the audience not only satisfied but elevated. Persuasion is about delivering a neural crescendo. Neuroscience tells us that endings shape memory: the hippocampus consolidates experiences more effectively when they conclude on a note of emotional clarity. But there is more: an emotional lift creates forward momentum.

Neurobiologically, this upswing corresponds to a surge in dopamine. When a narrative shifts from pain to resolution, especially one that hints at renewed possibility, the brain releases dopamine in anticipation of reward. This chemical signal not only reinforces the emotional content but also primes the brain for future action. It is as if the story closes not with a period, but with an arrow—pointing toward what is next. The primal brain is soothed, the limbic system resonates with closure, and the rational mind eagerly justifies the emotional journey it just endorsed. The bottom-up effect propels the next step in the decision-making process: rationalization.

Note that a well-crafted narrative does not necessarily end with certainty; instead, it ends with energized possibility. That closing emotional lift is what transforms a story from something we heard into something we believe—and act upon. The architecture, based on pain and three emotional claims — joyful expectation — is formulaic but also foundational. It reflects the sequence by which real change happens in the brain: from alarm, to engagement, to belief, to action. When we align our narratives with that primal sequence, we offer the brain what it has always craved: meaning, coherence, and a path forward.

We observe this pattern in daily life. Think about your last visit to the supermarket. What do you remember? The moment you walked in, you may have been greeted by the scent of fresh bread or a chaotic display of promotional signage. Moreover, what about when you left? The long line at the register, or the relief of getting a receipt and heading home. These bookends color your entire impression of the experience, though they represent just a fraction of the total time spent in the store.

Furthermore, in between, something subtle but neurologically potent may have happened: the cashier's demeanor. If the person behind the counter greeted you with a warm smile and genuine eye contact, your limbic brain registered a micro-dose of oxytocin—a neurochemical linked to trust and social bonding. Conversely, if you were treated like a number, rushed through the transaction without acknowledgment, your brain may have filed the encounter under "threat" or "inconvenience," tagging it with cortisol-fueled aversion.

In both cases, the emotional tonality of the moment disproportionately influenced your memory and impression of the entire shopping trip. This is emotional encoding in action: our brains do more than recall the facts of an experience. They distill their emotional valence, compressing it into a story that either invites return or triggers avoidance.

Alternatively, consider a movie. You can often recall the opening scene or the final plot twist, but struggle to remember the second act. First and last impressions act as anchors, distorting our memory of the in-between. Your memory is not necessarily failing—it is optimizing the blueprint of an experience. Why? Because the brain must be efficient.

Encoding every moment is metabolically costly and cognitively wasteful. Instead, the hippocampus—our brain's memory librarian—strategically bookmarks beginnings and endings as the most informative segments for future reference.

For communicators, this is liberating. You do not need to win every second of your audience's attention. You need to win the first one and the last.

Trigger 6: Familiarity

The *familiarity trigger* may seem to echo the *change trigger* as well, as it plays with recognition and contrast. However, the two operate like neural opposites in the brain's theater. While change commands attention through surprise, novelty, and disruption, Familiarity does the opposite: it earns trust and comfort through repetition, pattern recognition, and predictability.

Think of change as the unexpected sound that makes your primal brain whip its head around. Familiarity, by contrast, is the gentle hum of a melody you have heard a hundred times—the mental shorthand for "safe," "known," and "low cognitive effort." Where change jolts, familiarity soothes. Where change interrupts, familiarity flows.

For instance, we can recognize familiar faces with remarkable speed, often within 100 to 200 milliseconds, and usually complete recognition by about 400 milliseconds. Brain imaging studies reveal that specialized regions, such as the fusiform face area, rapidly trigger a "familiar face signal," enabling us to identify someone we know well before we are consciously aware of it. This ability is faster than recognizing most objects, reflecting the evolutionary importance of faces in social survival and connection.

In contrast, recognition of non-face objects, such as animals or tools, typically activates different cortical areas and occurs on a slightly slower timeline. For example, object categorization can begin as early as 150 milliseconds but usually stabilizes around 500 milliseconds. This timing depends on complexity and familiarity: highly familiar objects (like a household item) can be identified more quickly than novel ones. For living creatures, such as animals, the brain employs an intricate interplay between the lateral occipital cortex and the parahippocampal regions, which takes approximately 300 to 600 milliseconds for reliable classification.

Moreover, studies using neuroimaging tools like EEG and MEG show that the brain starts forming coarse-level pattern recognition—such as distinguishing a natural scene from an artificial one—in under 200 milliseconds. This is supported by the brain's propensity for "gist detection," a survival mechanism that enables us to assess visual scenes with minimal input rapidly. Imagine walking into a room and instantly sensing that it is a kitchen—or that a forest trail looks safe or threatening. That immediate, intuitive impression is the gist. It is your brain's way of saying, "I get the big picture"—and it usually happens in under 200 milliseconds.

This rapid gist perception is also about grasping the overall structure and emotional tone: familiar versus novel, natural versus artificial, safe versus threatening. From a neuromarketing perspective, gist is crucial. It is the brain's first filter in the visual hierarchy and heavily influences whether attention is sustained or abandoned. If the gist of a webpage, ad, or package feels confusing or irrelevant, the primal brain will often reject it before conscious thought kicks in. When designing persuasive content, the key takeaway is to optimize for an immediate, emotionally clear gist. The faster the brain "gets" what you are showing—and why it matters—the higher the chances your message will survive the primal filter. Together, these insights reaffirm that the brain's visual processing is neither linear nor uniform. It prioritizes evolutionarily significant stimuli, such as faces and threats, allowing them to bypass the processing queue. This makes recognition a dance of neural shortcuts and hierarchies—tuned more for speed and safety than for accuracy or detail.

In a world constantly competing for attention, marketers often over-index on novelty and forget that the brain craves safety and fluency. Processing familiar stimuli is neurologically less taxing and emotionally reassuring. In fact, the familiarity trigger is underpinned by what psychologists call the "mere-exposure effect," first demonstrated by Robert Zajonc (1968). In a series of experiments, Zajonc demonstrated that people consistently rated nonsense words, unfamiliar faces, and abstract images as more likable and trustworthy when they had been repeatedly exposed to them—even when the exposure was brief and unconscious. This effect occurs because repeated exposure lowers cognitive effort and threat perception, signaling to the primal brain that something familiar is safe.

Modern neuroscience confirms what Zajonc's 1968 mere-exposure experiments first hinted at: the brain is a pattern-recognition engine that rewards the familiar and flags the novel for scrutiny. *f*MRI research indicates that repeated exposure to a stimulus reduces neural activity in the lateral orbitofrontal cortex (LOFC), a region associated with negative affect and threat evaluation. This "repetition suppression" signals that the stimulus is no longer a threat, reducing emotional resistance and vigilance (Zald & Kim, 2002; Grill-Spector et al., 2006).

But this neural quieting is not a symptom of cognitive laziness; it is a hallmark of evolved efficiency. In fact, master chess players exhibit less brain activation than novices during gameplay, not because they think less, but because their pattern-recognition circuits have been optimized for speed and precision (Bilalić et al., 2011). The brain, like a seasoned athlete, conserves energy by running streamlined routines. Even Einstein's brain was found to exhibit unusually high myelination in key white matter tracts, suggesting that intellectual power often derives from interconnectedness *and* global efficiency, not just isolated cognitive fireworks (Witelson et al., 1999).

Thus, repeated exposure fosters similar optimization. For example, longitudinal *f*MRI studies have shown that stimuli exposed over ten days trigger a shift from novelty- and effort-detection regions to primal valuation networks, such as the ventral striatum (Reber et al., 2017). Familiarity, then, is not passive. It reflects the brain's drive to conserve cognitive resources while maximizing meaning. The brain becomes neurologically economical with what it knows. All these findings reinforce the persuasive value of branding consistency, auditory logos, and visual repetition—not as aesthetic choices, but as neurological shortcuts that lower resistance and build implicit trust. Familiarity does not just feel good; it frees cognitive bandwidth and whispers to the primal brain, "This is safe."

Marketers who understand this principle leverage it relentlessly. Consider the jingles of McDonald's, the Netflix "ta-dum" sound, or Coca-Cola's use of its signature red script. These are not arbitrary creative flourishes—they are consistent sensory signals that build implicit trust through repetition. Another compelling example is the Intel "bong"—a simple five-note mnemonic that has become synonymous with technological reliability. This auditory signature has been shown to activate emotional and memory-related brain circuits, even when presented without accompanying visuals, a fact I was able to confirm during a neuromarketing engagement with the legendary chip maker. Consumers not only recognize it instantly but often feel a sense of assurance the moment it plays. When such sonic cues are repeated consistently, they transcend branding and become cognitive shortcuts—neurological fast lanes that reduce effort, increase fluency, and anchor trust.

In the world of neuromarketing, this kind of hardwired familiarity is indispensable. Even political campaigns exploit this primal bias: studies have shown that candidates with names that sound more familiar tend to receive more votes, independent of their policies. Meanwhile, companies with ticker symbols that are easy to pronounce or congruent with their names tend to outperform those with less intuitive symbols. Memorable ticker symbols, such as HOG (Harley-Davidson) and FUN (Six Flags), attract more attention and investor interest. Thus, while the *change trigger* wakes the brain up, the *familiarity trigger* helps it relax and say yes. Achieving brand consistency through repeated visual cues and tonal alignment across platforms is not just a cosmetic choice, they all serve as a neural lubricant, reducing friction and making persuasion easier.

Unfortunately, many marketers fall into the trap of relentless reinvention—altering logos, jingles, or campaign themes not necessarily because consumers demand it, but because the advertising ecosystem rewards it. Agencies and media platforms often push for novelty to justify new budgets and creative cycles. Yet neuroscience confirms that the brain thrives on repetition and pattern stability to build trust. The Humane Society's long-running ad, featuring Sarah McLachlan's "Angel," is a case in point: unchanged for two decades, it has raised millions through emotional familiarity alone.

When I bring up these examples during my talks, people often object that some successful brands do not apply these principles and are constantly changing their ads. A good example is Geico. However, Geico's wildly diverse ads illustrate a deeper form of consistency: *reliable absurdity*. Though each campaign varies—talking geckos, cavemen, taglines about saving 15%—the *meta-pattern* is consistent: surprising, entertaining creativity delivered at high frequency. Over time, viewers come to expect and enjoy novelty itself, which becomes the familiar hook. This reflects an advanced persuasive strategy: establishing a consistent *tone and tempo*, even when the surface content changes. It is familiarity in disguise.

So, how can we ethically harness this powerful yet often overlooked trigger to meet the brain's deep craving for trust and familiarity? The answer lies in understanding the primal brain's appetite for the concrete, the demonstrable, and the verifiable. We are wired to trust what we see, touch, or sample—not out of habit, but because doing so has always been a matter of survival. Abstract promises rarely move the needle unless they are anchored in sensory reality. To persuade the primal brain, evidence must be delivered not as a future potential but as a present experience. Familiarity becomes a powerful primal trigger primarily when it is supported by tangible evidence.

But it is not just *any* evidence the primal brain wants; it demands that proof be delivered *in the present moment*. The primal brain is not designed to conceptualize the future with nuance; it cannot easily simulate delayed gratification or abstract gains. That is the domain of the rational brain. To persuade effectively, marketers must present evidence as something unfolding in the now: make the benefit visible, bring the transformation to life, and dramatize the loss being avoided today. The more immediate and tangible the experience is, the more convincingly it speaks to the primal brain.

Familiarity without immediacy can feel empty. When familiar cues are paired with sensory-rich, present-tense proof, the primal brain sees not just a safe option—but a safe option it can act on right away. Neuroscience confirms that sensory experiences and physical interactions create deeper encoding pathways in the brain. When people engage with physical samples or observe demonstrations, they activate multimodal areas of the cortex that go far beyond passive reception. The brain is not just watching, it is simulating, anticipating, and emotionally indexing the experience. This neural engagement leads to stronger memory traces, heightened emotional resonance, and increased confidence in decision-making (Schifferstein, 2006). Our craving for this kind of experiential evidence is so primal that it often overrides logic.

Think about how often you touch a hot plate after being warned not to—to feel if it is *really* hot. Alternatively, the near-universal reflex of dipping a toe into the pool before jumping in, even when the temperature is clearly posted. These irrational behaviors defy logic, but they reflect the primal brain's need for immediate, embodied evidence. We crave real-time sampling because our survival has always depended on it. Even kissing, in its evolutionary origins, is more than affection—it is a biological sampling tactic deeply anchored in our primal wiring. Research shows that kissing facilitates chemical and tactile assessment of compatibility. Saliva carries immune markers, hormonal cues, and genetic signals, which help gauge mate fitness. One study found that kissing ability strongly influenced desirability during courtship, especially for women (Wlodarski & Dunbar, 2015). Another proposed that kissing evolved from primate grooming behaviors, where lip contact marked social trust and bonding (Lameira, 2024).

This sensory logic extends to scent. In a seminal study, women smelled T-shirts worn by different men and consistently preferred those whose immune systems were most complementary to their own (Wedekind et al., 1995). The lesson? Trust is not built on abstract claims. It is earned through palpable, immediate cues. In marketing, this is reflected in the enduring power of in-store demonstrations, test drives, and interactive product previews. Seeing is believing—but touching is trusting. If physical sampling is not possible, marketers must simulate the experience through vivid, sensory-rich language or videos. Familiarity, when paired with sensory immediacy, does not just trigger recognition. It creates visceral endorsement—the kind the primal brain remembers and acts upon.

For instance, Do-It-Yourself (DIY) videos that show hands performing tasks—such as cutting, assembling, and painting—are not just visually satisfying; they are also cognitively superior for establishing primal evidence. A 2024 study found that instructional videos featuring visible hand drawing or demonstrating actions significantly enhanced viewer comprehension compared to static images or cursor-only interactions. This ties into the *enactment effect,* a well-established phenomenon in cognitive psychology where memory is improved when people observe or perform meaningful actions. Hands, primarily when actively engaged, anchor learning by adding tangibility and movement, which the brain finds easier to encode and recall.

This is especially true for procedural knowledge—learning how to do something—where visual demonstrations outperform abstract explanations. Watching someone's hands is more than just a visual cue; it becomes a form of embodied cognition, where the viewer's brain simulates the action. For marketers and content creators, this insight is gold: showing the product in use, in context, makes the information stick. In neuromarketing terms, it is a tangible trigger that leverages both primal relevance and cognitive fluency to boost engagement, retention, and conversion.

Unfortunately, many marketers fail to grasp this fundamental principle of human behavior. They rely on vague promises, generic images, or abstract concepts, failing to understand that the primal brain does not respond to ambiguity. It wants evidence. Show me, do not tell me. Let me sample, touch, and taste. Familiarity is not just comfort; it is proof. Moreover, in the hierarchy of persuasion, proof outranks vision statements, especially when paired with sensory information (Kousta et al., 2011). Instead of saying "we optimize operational efficiency," say "we speed up your workflow, so your team gets home before dark." Abstract ideas are the enemy of the primal brain. We evolved to navigate forests, not spreadsheets. Concrete language—things we can see, touch, smell, or feel—is more persuasive and easier to remember. Metaphors help translate complexity into brain-friendly formats.

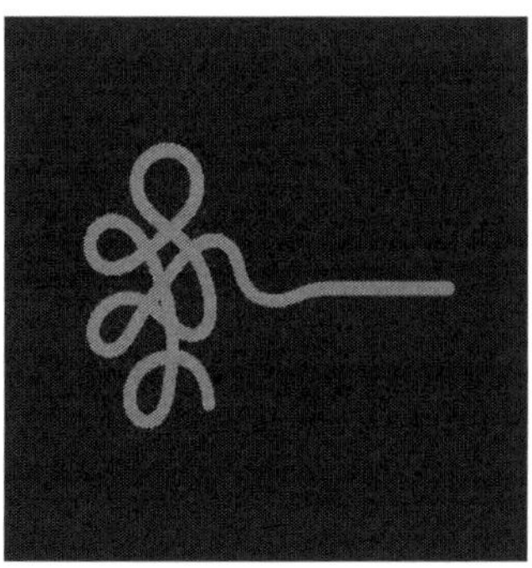

Trigger 7: Simplicity

If there is one undeniable truth about the human brain, it is that it was not built for the world we live in now. Despite the exponential growth of information in our environment, the brain's processing capacity has remained static. Neuroscientific research estimates that while our senses take in a billion bits of information per second, our conscious cognitive system can handle only about 10 bits per second (Leshin, 2024). That is one word per second—on a good day.

Now consider this: the average person today is exposed to more information in 24 hours than someone in 1900 encountered in their entire lifetime (Bawden & Robinson, 2009). Whether or not that is an exact numerical match, the implication is clear—the volume, complexity, and velocity of modern communication dramatically outpace our ability to process it.

This mismatch is neurologically exhausting. Every extra layer of complexity—an ambiguous phrase, a dense layout, a new term—

demands more energy from the brain's limited attentional system. Rational regions, such as the anterior cingulate cortex, which monitor cognitive conflict and effort, become more active as the brain struggles to interpret and resolve ambiguity. The default response, especially from the primal brain, is to conserve energy by checking out.

Simplicity, then, is a survival strategy for communication. In an environment where attention is scarce and mental effort is costly, messages that are clear, intuitive, and easy to digest enjoy a massive competitive advantage. The *simplicity trigger* works not because it reduces complexity, but because it respects the biological limits of the human decision-making system. If your message cannot be grasped within the brain's narrow bandwidth, it may never be processed at all—no matter how important or clever it is.

The bottom line is that cognitive overload leads to dropouts. Neither are our brains designed to multitask deeply or process high volumes of abstract data without friction. Processing fluency, the ease with which information is understood, has been linked to greater perceived truth, likability, and even aesthetic pleasure (Reber et al., 2004). When the brain processes something effortlessly, it makes mistakes that are easy to correct. Simplicity, therefore, is about facilitating fluency (Alter & Oppenheimer, 2009).

It is tempting to frame complexity as the opposite of simplicity, but that is a misleading dichotomy. Both "simple" and "complex" are subjective judgments, heavily influenced by factors such as age, education, cultural context, cognitive capacity, and even emotional state. What is simple to a neuroscientist might feel impenetrable to a teenager, and what is intuitive to one culture might seem chaotic to another. Rather than getting tangled in definitions, it is more helpful to focus on what complexity does to the brain.

In my research and numerous neuromarketing studies, I have observed that when people encounter complex messaging, such as jargon, heavy language, ambiguous visuals, or layered choices, the brain's conflict monitoring system is activated. Specifically, the anterior cingulate cortex (ACC) becomes more active (Botvinick et al., 2004), signaling a type of neural warning. The ACC flags the discrepancy between what is expected and what is being presented, leading to cognitive strain. In response, the brain often slows down processing, recruits more effort, and—if the strain persists—may disengage altogether.

On the other hand, when people encounter clear, familiar, and structured stimuli, the brain shifts into a more receptive mode. Predictable inputs tend to activate reward-related areas, such as the ventral striatum, which are associated with ease, fluency, and even pleasure. Rather than conflict, the brain increases engagement and prompts further action.

The takeaway is not that all marketing must be dumbed down. It is that every layer of complexity introduced into your message adds friction. Unless there is a strong, motivating reason for the brain to push through that friction, it will not. It will just move on.

Again, this ties directly to the dual-process theory widely popularized by Daniel Kahneman but initially proposed by cognitive scientists such as Keith Stanovich and Richard West. While Kahneman masterfully articulated the behavioral aspects of these two systems in his book *Thinking, Fast and Slow*, it is essential to acknowledge that his work relied heavily on psychological modeling and behavioral economics rather than detailed anatomical neuroscience.

In fact, the neural underpinnings of what Daniel Kahneman calls System 1 and System 2 remain loosely mapped in the literature. Therefore, while they provide a useful mental model for distinguishing between fast and slow thinking, the actual biology is more of a continuum than a switch.

Through my neuromarketing work, I have observed that when messages are simple and emotionally resonant, they tend to activate the primal brain's rapid-response circuits, particularly regions involved in associative memory (like the hippocampus), pattern recognition (often involving the basal ganglia), and emotional processing (such as the amygdala and ventromedial prefrontal cortex). This is what we typically label as "intuition." But intuition is not magic. It is not some ethereal sixth sense. It is a product of compressed experience, formed through neural shortcuts that develop over time through repetition, prediction, and emotionally salient feedback.

From a neuroscience standpoint, intuitive thinking is a form of highly efficient inference. The brain is not thinking less; it is thinking faster by skipping deliberation and jumping straight to a conclusion based on previous exposure. That is why intuitive judgments feel effortless and automatic. Basically, since the heavy cognitive lifting happened in the past, those pathways are now well-myelinated highways.

But introduce complexity—when even a single element in the message demands extra interpretation, such as unfamiliar phrasing, abstract visuals, or unexpected structure—and the brain's predictive shortcuts begin to break down. The primal brain, wired to conserve energy and prioritize clarity, now faces a costlier task: it must slow down, allocate more cognitive resources, and resolve ambiguity before deciding what to do next. This forces a shift to slower, more resource-intensive reasoning associated with System 2. Areas such as the dorsolateral prefrontal cortex and anterior cingulate cortex become active as the brain engages in problem-solving, error monitoring, and deliberate evaluation.

Here is the catch: the primal brain, which is metabolically conservative and attention-averse, often shuts down long before the rational brain has a chance to weigh in. If the intuitive system does not receive or accept it, the rational system is rarely summoned at all. So, in a world saturated with noise and limited attention, persuasion is not just about accuracy or logic. It is about making messages feel intuitive—not because they are dumbed down, but because they have been designed to meet the brain where it is fastest, most efficient, and most emotionally attuned. Think about successful product design. Apple does not boast "multi-core dynamic processing integration." It says, "Think different." IKEA diagrams use pictograms, not paragraphs. Nike has used its iconic slogan "Just Do It" for 37 years. McDonald's has used "I'm lovin' it since 2003. Also, the most persuasive political slogans are rarely long: "Yes We Can". Simple is sticky.

In a world drowning in information but starved for attention, simplicity is not a nice-to-have—it is non-negotiable. The human brain is constantly triaging what to notice, what to process, and what to ignore. Moreover, it often chooses the path of least resistance.

This is why simplicity is so powerful—and so dangerous at the same time. Simple messages move faster, stick longer, and get shared more widely. They sneak past the brain's filters and settle in deep before our slower, more rational systems can evaluate them. That is why clarity beats cleverness. Why, unfortunately, lies often outpace the truth. Simplicity is not always a virtue, but also a vulnerability. The same brain that craves clarity and speed can also be swayed by flawed logic when it is presented well. This is referred to as gullibility.

Gullibility thrives in environments where simplicity overrides scrutiny. The brain, especially its primal regions, is wired to conserve energy and prefer shortcuts—so it often defaults to intuitive, effortless interpretations rather than engaging in complex analysis. That is why simple, clear messages, even when misleading, can be more persuasive than nuanced truths. In persuasion science, simplicity is a cognitive vulnerability that skilled communicators can exploit.

This helps explain the appeal of conspiracy theories. They often reduce complexity into emotionally charged cause-and-effect narratives. They offer a villain, a hero, and a simple explanation for a complex world. They cater to our confirmation bias (we seek info that fits what we already believe). They exploit the proportionality bias (we believe big events must have significant causes). They satisfy our agency's bias (we assume things happen because someone intended them to).

From a neuromarketing standpoint, these are the dark side of the simplicity trigger. The message spreads because it fits the brain's low-energy, high-certainty preference—not because it is true. So let me help you recognize true simplicity in messaging, how to engineer it deliberately, and how to avoid the trap of oversimplification that fuels bias, confusion, and the spread of misinformation.

Use the following questions to assess the simplicity of your message—not based on your intention as the sender, but from the perspective of the human brain as the receiver:

1. **Can it be understood in under 3 seconds?**

If the core idea is not graspable within a few seconds, it triggers cognitive strain. Test: Say it aloud. If it requires qualifiers, it is too dense.

2. **Does it pass the "explain it to a 10-year-old" test?**

Strip it down to nouns and verbs. Would a child understand the core point without additional context? Does it feel like you are dumbing down the message or ridding it of unnecessary cognitive burden?

3. **Is it structured with a clear beginning, middle, and end?**

Humans remember stories, not bullet points. Simplicity often hides in structure: make the flow intuitive, not fragmented.

4. **Is the vocabulary familiar?**

Big words and jargon may impress you, but they trigger effort and confusion in others. Tip: Choose words your audience already uses. And get rid of acronyms!

5. **Does it feel instantly authentic or emotionally resonant?**

The primal brain favors fluency and feeling over analysis. Simplicity often rides on emotional shorthand.

6. **Is the format scannable?**

Use visual hierarchy, spacing, and brevity. The eye should "get it" before the brain reads it.

Here are more ways to convert complex messages into simple ones:

1. **Distill the essence**: Do so without losing the thread of your story. Identify the "essential question" your message answers. Remove everything that does not directly support that answer.

2. **Chunk it**: Break ideas into digestible parts: 3s and 5s work best, and use modular thinking: one idea per slide, line, or beat.

3. **Prototype for speed**: Use tools like elevator pitches, 1-line summaries, or sticky note tests. If your team cannot repeat it the same way twice, it is not simple yet.

4. **Test for Recall**: After sharing your message, ask someone to repeat it 10 minutes later. If it sticks, it is simple.

5. **Measure what gets shared**. Viral messages are often proxies for simplicity.

6. **Use the dinner party test:** Would someone say this aloud at a dinner party? Would they text it? Would they post it?

Make your message a meme.

A *meme* is a compressed cognitive unit. Essentially, it is a small, easily digestible idea that spreads because it requires minimal mental effort to process and remember. *Memes* deserve more of our attention as we close on the simplicity trigger. They function as neurological shortcuts. They pack meaning into tight, often humorous or emotionally charged formats (such as images, phrases, or slogans) that the primal brain can instantly recognize and act upon. From a neuromarketing perspective, memes speak directly to the primal brain, aligning perfectly with its demand for cognitive fluency.

Memes are powerful vehicles because they deliver compressed, emotionally charged ideas that the brain processes almost instantly. Take "Keep Calm and Carry On"—five words that soothe, instruct, and stick. It is simple, rhythmic, and adaptable. The "Distracted Boyfriend" meme captures desire, betrayal, and conflict in one frozen moment—no caption needed. "This Is Fine," with a dog calmly sipping coffee in a burning room, speaks volumes about denial and modern anxiety. Grumpy Cat's scowl became a global symbol of cranky indifference with just one image. "One Does Not Simply Walk Into Mordor" dramatizes complexity using an iconic still and structured phrase. Likewise, SpongeBob's "Ight, Imma Head Out" is pure emotional shorthand for discomfort and escape.

What unites these memes is their cognitive fluency—they are fast, frictionless, and deeply resonant. They bypass logic and speak straight to the primal brain, often using humor, contrast, or familiar emotional templates. In persuasion, that is gold. Memes demonstrate that complexity is not necessary for impact; simplicity, paired with emotion, is enough to encode ideas into memory and spark viral momentum. They are not just internet culture—they are modern hieroglyphs for the instinctive mind. When used strategically, memes become tools of influence as potent as any headline or slogan.

To wrap up, in the battle for attention, simplicity is your sharpest tool. It speeds up understanding, lowers cognitive resistance, and increases the odds your message will be remembered—and repeated. Whether you are crafting a campaign or prompting an AI, clarity is not optional. It is your first filter.

Now that we have cleared the path for understanding, it is time to explore how people decide what to trust your messages—especially when they are unsure. Enter the Social Proof Trigger.

Trigger 8: Social Proof

In my years of research, few phenomena have demonstrated such consistent power over decision-making as social proof. It is not simply that we imitate others; instead, our brains are wired to equate consensus with correctness. Especially when we are unsure or under pressure, the actions of others become our behavioral compass.

However, social proof is often misused. In countless campaigns, it is reduced to a cascade of star ratings, baked quotes with questionable origins, or generic statistics. These tactics may satisfy a marketing checklist, but they often miss the primal brain. My research over the last 20 years, through both private sector work and academic projects, has demonstrated that social proof only becomes truly persuasive when it is formatted for the human brain. This means activating the right emotional and cognitive systems—not just displaying data, but orchestrating experiences that whisper to the primal brain, "Hey, you are not alone."

Consider customer testimonials that resonate with fears and hopes, immersive demos displaying others' real-time reactions, or behind-the-scenes glimpses that humanize a brand. The most persuasive social proof is more than sheer ratings; it is about achieving biological resonance. Real emotions—like a quiver in the voice or a tearful smile—carry weight because our brains tag them as credible. These trust signals deliver visual and emotional evidence in milliseconds.

The neuroscience behind social proof is contagious. When we witness someone emotionally endorsing a product, we are not just spectators. Mirror neurons, first discovered in the 1990s by Giacomo Rizzolatti and colleagues, fire as if we are experiencing the same emotion ourselves. It is empathy at a neural level (Rizzolatti, G., et al, 1996). The brain simulates what it observes, turning another's experience into a simulated trial of belief.

Social validation also activates areas involved in trust and reward, like the medial prefrontal cortex and ventral striatum (Klucharev et al., 2009). When we see a relatable person endorse a product, it generates a neurological pat on the back. These mechanisms—mirror neurons and reward circuits—form the scaffold of persuasive social proof. They explain why emotional testimonials outperform statistics. One video of a joyful customer outweighs thousands of numerical ratings in its ability to move the primal brain.

In our lab, I have observed that testimonial-driven campaigns consistently outperform feature-driven ones, particularly in high-stakes decisions such as healthcare, insurance, or home buying. These are contexts where the brain seeks clarity and safety. In those moments, testimonials act as neural shortcuts to trust. Moreover, the more emotionally genuine the endorsement, the more persuasive it becomes.

This effect stems from our innate tendency towards conformity—a survival mechanism that enables us to align with group norms in uncertain situations. The classic Solomon-Asch experiments of the 1950s demonstrated this beautifully: participants often agreed with obviously incorrect group answers to avoid social dissonance. The takeaway? Social pressure can override one's own perception (Asch, 1951).

Figure 21: The Solomon Ash Experiment

Neuromarketing applies this insight not to manipulate, but to reassure. Our brains lean toward what others seem to trust, especially when personal risk is high. The more relatable the source, the stronger its influence. Neuroimaging confirms that similarity in endorsements (e.g., mothers endorsing for other mothers) activates reward areas more strongly (Zaki et al., 2011).

Meanwhile, not all social proof is equal. For instance, celebrity endorsements, although attention-grabbing, offer modest persuasive power, with an average 10% gain (Erdogan, 1999; Amos et al., 2008). Why? Because they often lack perceived credibility and personal relevance. Despite the outrageous sums spent by advertisers to feature celebrities in commercials, the return on investment is dubious at best. The primal brain, concerned with safety and trust, is not easily impressed by status alone, especially when the endorsement feels scripted. In some cases, it may even backfire. When celebrities appear in ads that feel disconnected from the product or overly scripted, viewers can become skeptical or even resistant. Research has shown that incongruent endorsements—where there is a mismatch between celebrity and product—can reduce ad credibility and negatively impact brand attitudes (Till & Busler, 2000).

So, what deepens the impact of social proof beyond testimony? Demonstrations offer the next evolutionary step—from imitating others to briefly becoming them. This is the transportation effect at work. It is no longer about observing consensus; it is about immersing oneself in the experience of that consensus. Demonstrations enable our mirror neuron system to simulate not only approval, but also participation. In those moments, belief is not a hypothetical concept; it is a visceral experience. For the primal brain, that is the ultimate proof of safety.

To deepen persuasion, we must embed people in mini-experiences that anchor belief—not just that others approve, but that the product or solution works for them too. Watching someone discover value with genuine emotion triggers mirror neurons, allowing the viewer to simulate that experience. The brain registers not just approval, but projected ownership. This is where demonstrations shine.

Demonstrations bridge the gap between social validation and sensory experience. They allow the primal brain to experience the benefit, even if only briefly. In the NeuroMap® model, demos are "meaningful trust bites, "mini-experiences in the only timeline the primal brain understands. We are biologically wired for this. It explains why we sample food before buying it, touch a hot plate despite warnings, and test intimacy with a first kiss. As we discussed in the familiarity trigger section, kissing is thought to function as a chemical sampling system—a biological assay of compatibility. This simple act is a sophisticated sensory test embedded in our mating strategies, another primal demonstration designed to predict future reward or risk. All these behaviors serve as real-time validations of safety and reward.

When done right, demonstrations offer fast-tracked trust. They tell the primal brain, "This feels real. We are safe." Marketers should treat demos as neurological necessities—not decorative fluff. In the architecture of persuasion, they are not window dressing. They are the cornerstone of belief.

As we have seen, social proof is compelling because it alleviates the brain's uncertainty and leverages our innate need to conform safely. Demonstrations amplify this effect by allowing the brain to experience trust—instantly and viscerally. However, not all persuasive tools adhere to these rules of comfort. Sometimes, the fastest path to attention is not through reassurance, but rather through interruption. This brings us to our next primal trigger: disruption.

Trigger 9: Disruption

While both change and disruption activate attention in the brain, they are not processed the same way. Change can be incremental, familiar, and often expected—gradually handled by the brain's predictive coding system in the neocortex, which constantly updates mental models to reduce uncertainty. Disruption, however, is categorically different for the brain. It signals a potential threat or opportunity significant enough to override routine processing. This first alerts the primal brain, particularly the amygdala, triggering a heightened state of arousal. Disruption hijacks attention by violating expectations—provoking surprise, confusion, or even fear. Neurobiologically, it is closer to a fight-or-flight cue than a gentle nudge. For persuasion to work, especially in saturated markets, messages must disrupt—not merely change—the way the brain frames a brand or problem.

Disruption is a diagnostic trigger for the primal brain. One of the most illustrative examples comes from a research technique we will explore later in the book: the Implicit Association Test (IAT). This method asks participants to quickly sort concepts—like "trustworthy" or “risky”—into categories. But here is where it gets fascinating. By suddenly flashing an unexpected image mid-task, a snarling dog, a crying child, or a political figure, the test disrupts the brain’s automatic responses. Reaction times are slow. Errors spike. Attention sharpens.

This small jolt reveals something powerful: disruption forces the brain out of autopilot. It exposes biases and latent associations that we are not aware of. In persuasion, the same principle applies. When we break the expected pattern, we interrupt the habituation process. Furthermore, when the brain can no longer predict what is coming next, it starts paying remarkably close attention. That is disruption, a tool for revelation and re-engagement.

In the grand theater of attention, nothing grabs a seat faster than the unexpected. The brain is a prediction machine, constantly anticipating the next beat in its sensory symphony. When those expectations are violated, the brain does not retreat—it leans in. That moment of disruption, whether it is a sound, a color, or a narrative twist, is a neural fire alarm.

This kind of unexpected stimulus activates not only the locus coeruleus (LC), the brain’s primary factory for norepinephrine, a neurotransmitter crucial for alertness and vigilance, but also the amygdala, which acts as the brain’s emotional megaphone. Together, they snap the brain out of autopilot and prepare it to respond to potential threats or novel opportunities. Remember that the amygdala does not whisper; it shouts. When a stimulus deviates from expectations, the amygdala assesses the signal for potential threat or reward. That jolt of arousal is then paired with a rush of norepinephrine, which sharpens attention, enhances encoding, and boosts the likelihood of memory formation (Sara, 2009).

From an evolutionary perspective, this system was our early warning mechanism under the fierce stewardship of the primal brain. If a shadow moved unexpectedly, it might have been a predator. Today, it is more likely to be a pop-up ad or a meme—but the primal brain responds with the same alertness.

In neuromarketing, disruption cannot be about gimmicks. It is about the kind of meaningful, biologically relevant surprise that the brain can recognize and react to. Not all novelty registers as disruption. When something is too unfamiliar—like those legendary tales of Indigenous people allegedly failing to see European ships because they had no conceptual framework for them—it may not even register perceptually.

True disruption, then, requires a delicate balance: it must be different enough to stand out, but not so alien that the brain discards it as noise.

Recently, I experienced this firsthand. I was flying from Kauai to the Big Island, believing it was a direct flight. But as the plane began its descent, I glanced out the window and immediately saw a surprising landscape and clusters of houses I had never seen before. Despite the visual evidence, including tall buildings and a layout I was familiar with from living in Honolulu before moving to Kona, I was still convinced we were landing on the Big Island. My brain was so anchored in the original belief that it filtered out what my eyes were telling me. That is the power of expectation.

Disruption works because it challenges these predictive models. But to be effective, it must land within the boundaries of what the brain can still process and revise—not so foreign that it has been rejected outright, and not so subtle that it goes unnoticed. When it resonates with the primal brain, the message often taps into deeply ingrained sensory, emotional, or pattern-based expectations. Break a rhythm. Flip a script. Deliver a punchline where a platitude was expected. These are the types of shifts that interrupt the brain's autopilot without triggering confusion. Think of a familiar story twisted with an ironic turn. This landing page breaks visual monotony with a sudden color or layout inversion, or a video testimonial delivered in reverse chronology. These moments activate curiosity—the brain's most underrated motivational force.

Iconic campaigns have mastered this. Consider Apple's "1984" Super Bowl ad, which broke every rule of commercial storytelling with its dystopian imagery and Orwellian tone. How about Old Spice's "The Man Your Man Could Smell Like"—a campaign that disrupted traditional male grooming ads with absurdity and fast-paced narrative jumps. These were not just clever. They were carefully engineered "neurointerruptions" that hijacked attention and refreshed brand perception.

In film, Christopher Nolan's "Memento" and Bong Joon-ho's "Parasite" use narrative structure as a disruptive device. Their storylines resist predictability, which forces the brain to stay engaged, to resolve what it cannot immediately categorize. This is why disruption is not merely a matter of surprise. It is managed unpredictability. It demands the brain build a new map—and in doing so, it deepens memory encoding.

But there is a delicate balance. Too much disruption without a payoff breeds distrust. It is like serving chili at a wine tasting—memorable, yes, but for the wrong reasons. In my campaigns, spanning retail to tech, strategic narrative disruption has consistently elevated engagement. Why? Because it does not just break the pattern—it offers a new one. It opens a neural window, and if what follows is valuable, the brain rewards the source with attention and recall.

In a world of sameness, disruption is currency. When deployed with purpose and timing, it not only grabs attention—it earns it.

Trigger 10: Anticipation

Anticipation is the spark that keeps the brain on edge, hovering between curiosity and reward. It is a uniquely human state of tension and focus, bridging the gap between the present moment and a desired outcome. From a neuromarketing perspective, anticipation is about crafting cognitive tension that holds the brain in a state of motivated attention. It is essential to clarify that cognitive tension is distinct from cognitive friction, which we will examine in detail in the last section on neuro-flow. Friction signals confusion or overload and prompts withdrawal. Tension, in contrast, sustains attention by leveraging the brain's reward circuitry to seek resolution. When done right, anticipation does more than fuel interest; it generates a biological urge to resolve uncertainty. This makes it a vital primal trigger.

While the primal brain lives in the present, scanning for immediate relevance and survival cues, the rational brain is built to navigate time. The human prefrontal cortex (PFC), particularly the dorsolateral prefrontal cortex, excels at reflecting on the past and simulating future outcomes. This capacity to project forward, to imagine and rehearse what lies ahead, is what makes anticipation such a powerful lever in marketing and persuasion.

Yet it is essential to understand that this forward simulation begins with bottom-up signals. The primal brain, grounded in the present, anchors perception in immediate survival relevance. Only once this relevance is registered does the rational brain engage, using these inputs to simulate, forecast, and strategize. In this context, it is essential to distinguish the moment-by-moment action of neurotransmitters. Dopamine is central to anticipation; it surges in response to the expectation of reward, not just its delivery.

In persuasion, a reward is more than a good treat; it is the psychological carrot that convinces the brain a particular choice is worth making. Persuasion science and neuromarketing reveal that rewards manifest in various tangible forms. One type is the relief of pain, since the brain is wired to detect and avoid threats first. In this case, the reward comes from removing something unpleasant, such as an insurance ad that promises peace of mind, a health product that reduces discomfort, or a service guarantee that lowers the fear of loss. Here, the valid reward is safety and certainty.

Another form of reward is tangible gains, which are measurable benefits that improve life in practical ways, such as saving time or money, enhancing performance with a faster car or sharper tool, or enjoying greater convenience, like 24-hour delivery. These appeal to rational processing, but they are most persuasive when framed in simple, primal language that makes the benefit immediate and clear.

Persuasion also works through social and emotional rewards, which do not depend on material objects but rather on feelings and one's sense of identity. Belonging, status, attraction, approval, or admiration all activate the limbic system, which is central to emotion and memory, and these rewards are potent drivers of brand loyalty.

A final and especially persuasive type is anticipatory reward, which thrives on dopamine-driven anticipation. Teaser campaigns build excitement for what is

to come, scarcity messaging creates urgency with limited availability, and gamified systems spark motivation through points, levels, or unlockable benefits. In these cases, the real persuasive power lies not in the product itself but in the promise of something just out of reach. In this sense, a reward in persuasion can be defined as the perceived reduction of pain or the promise of gain that captures attention and mobilizes the primal brain to take action to get what we want!

Surprisingly, studies have shown that dopamine levels can spike higher before the reward is delivered than when the reward is received (Knutson et al., 2001). This makes dopamine a pleasure-driven, present-centric motivator that mobilizes attention and action.

In contrast, serotonin plays a different, complementary role. Often associated with mood regulation and long-term well-being, serotonin supports future-oriented behavior. It is involved in patience, delayed gratification, and long-range planning. That is why neuroscientists describe dopamine as the neurotransmitter of "wanting" or “pleasure” and serotonin as the chemistry of "contentment" or "happiness" (Berridge & Robinson, 1998; Dayan & Huys, 2009; Kranz et al., 2010). Together, they illustrate the layered time horizon of brain motivation: dopamine drives us toward what we anticipate now; serotonin helps us invest in what we value over time.

Marketers who understand this principle know how to build narrative tension, pace the delivery of benefits, and leave just enough questions unanswered to compel action. This is the psychology behind teaser campaigns, product countdowns, cliffhanger storytelling, and mystery offers. They do not just inform; they provoke.

Consider the success of pre-launch campaigns. When a tech brand leaks a cryptic image or announces an upcoming reveal date without specifics, the collective curiosity builds. Forums buzz, predictions fly, and consumers become emotionally invested in a product they have not even seen. What is at work here is not just curiosity but anticipatory engagement. The brain begins to simulate ownership, utility, and social impact. Moreover, once the reveal finally happens, the perceived value of the product is magnified by the tension that preceded it.

This phenomenon can be traced back to our evolutionary wiring. In environments of scarcity, anticipating a potential reward—like a food source or shelter—conferred a survival advantage. Being alert and mobilized before the reward appeared allowed early humans to act quickly and decisively. Today, this wiring remains intact. Anticipation keeps us neurologically invested in what comes next.

A striking example of anticipation's power can be found in the world of entertainment. Netflix, for instance, discovered that releasing entire seasons of a show at once reduces the suspense-driven engagement that weekly episode releases create. The anticipation gap disappeared, and with it, some of the social buzz. When "Game of Thrones" aired weekly, the cultural anticipation between episodes fueled conversation, speculation, and a deeper emotional investment. Anticipation extended the life and emotional relevance of the content.

In neuromarketing, anticipation is about strategic pacing. Imagine a website that gradually reveals a benefit as you scroll, or an email campaign that subtly hints at a surprise to be shown in the following message. These techniques align with the brain's love of completion. The *Zeigarnik Effect*—a psychological phenomenon first observed by Bluma Zeigarnik in the 1920s, later replicated in experimental studies (Ovsiankina, 1928; Baumeister & Masicampo, 2010)—demonstrates that people tend to remember uncompleted tasks more effectively than completed ones. Leaving things unresolved keeps the brain engaged.

But anticipation is not limited to promotional timing or content pacing. It is also deeply embedded in storytelling. A well-structured narrative creates tension, raises stakes, and keeps the brain hooked by constantly asking, "What happens next?" Stories that introduce a protagonist, a problem, and a promise of resolution activate not only emotional engagement but also mirror neuron systems. The brain begins to live the story, projecting itself forward into the narrative arc. This effect, which we have already discussed, is known as narrative transportation, and it significantly enhances persuasion by reducing counterarguing and increasing emotional resonance (Green & Brock, 2000).

Stories are essentially anticipation engines. They generate an emotional loop that pulls the brain forward. When marketers harness storytelling strategically—in video testimonials, landing page copy, product use cases—they create more than interest. They create an anticipatory rhythm that drives attention and action.

Significantly, anticipation also boosts memory. The hippocampus, involved in memory formation, is activated more strongly when people anticipate emotionally charged events. This means that campaigns designed with anticipation not only get noticed but are more likely to be remembered. Moreover, in a noisy digital environment, what is remembered often wins.

Now, anticipation must be handled with care. Overuse or manipulation can backfire. When a promise is made but not delivered, when the dopamine spike is followed by disappointment rather than reward, trust erodes. The primal brain, constantly scanning for pattern integrity, interprets broken expectations as threats and deception. This is also at the heart of how many addictive behaviors form. As extensively discussed in my previous books (The Serenity Code, OPEN, and OPEN

UP), the mechanism of anticipation—especially the dopamine-driven expectation loop—is front and center in most addictions. Anticipation hijacks motivational circuitry, reinforcing the craving loop even in the absence of an actual reward.

This also explains why many academics and researchers have raised alarms about media platforms, video games, and digital content strategies that exploit these mechanisms in toxic ways. Studies have shown that intermittent rewards, variable reinforcement schedules, and unpredictable feedback loops can condition compulsive engagement behaviors (Foddy & Savulescu, 2010; Montag et al., 2019; Turel et al., 2020). Incessant behaviors such as Instagram scrolling, chatting, texting, and refreshing email exemplify how these digital habits can morph into uncontrollable cravings, continuously reinforced by the dopaminergic system. These are not merely lifestyle quirks; they are neurobiologically driven behaviors shaped by anticipation itself. Therefore, the art of triggering anticipation is more than creating tension; it is about resolving it responsibly and respectfully.

This is where demonstrations and social proof can also play a complementary role. Once you have built up anticipation, delivering the payoff with credible and emotionally resonant proof or a multisensory experience allows the primal brain to shift from curiosity to confidence. It closes the loop. Marketers should think of anticipation as a psychological inhalation—a breath held in suspense—that must be released with resolution, not collapse.

In some cases, even the delay itself becomes part of the pleasure. This is particularly evident in luxury branding. High-end product experiences often delay gratification through elaborate packaging, multi-phase unboxing rituals, or exclusive access queues. These are not forms of inefficiencies; they are deliberate anticipation builders. The longer the wait, the greater the perceived value.

Even a pricing strategy can tap into anticipation. Limited-time offers, countdown timers, and early-bird discounts all rely on urgency and projected scarcity to stimulate forward-focused attention. The brain does not want to miss out on future gains, so it acts preemptively to secure that outcome.

To summarize, anticipation is one of the most neurologically rich triggers in the neuromarketing toolkit. It fuels curiosity, extends engagement, enhances memory, and amplifies emotional payoff. It draws the brain forward toward a promise, stimulating attention systems and reward circuitry before the actual reward is even present. When skillfully integrated into storytelling, product design, marketing cadence, and sensory experiences, anticipation becomes a powerful ally of persuasion.

If anticipation teases the brain with what is to come, engagement rewards it for staying. It transforms suspended attention into sustained focus, pulling the audience deeper into experience through curiosity, emotional resonance, or sensory involvement.

Trigger 11: Engagement

Engagement is the process of embedding thought, emotion, and motivation into a single cognitive stream. In neuromarketing, we define engagement as the dynamic neural alignment between content and audience. It is a sweet spot where messages capture attention over time, despite competing distractions. Engagement is where persuasion begins to take root.

From a neurological perspective, engagement begins in the primal brain's attentional system, which acts as the first gatekeeper for incoming stimuli. Initial attention is captured through bottom-up signals managed by primal structures, such as the thalamus and superior colliculus, ensuring that anything novel or emotionally salient grabs our perceptual spotlight. From there, engagement becomes a distributed phenomenon, involving attentional networks in the parietal cortex, reward systems in the striatum, and emotional processing centers in the amygdala. But most critically, the prefrontal cortex (rational brain) helps sustain engagement by regulating goal-directed focus. The more emotionally resonant and cognitively coherent the stimulus is, the more likely it is to recruit and sustain neural resources.

One of the most direct ways to measure engagement in neuroscience is through a method known as frontal asymmetry analysis, which utilizes EEG (electroencephalography). This technique tracks differences in electrical activity between the left and right frontal lobes. We will explore EEG and other neuromarketing research methods in greater depth later in Chapter 6.

Research has shown that greater activation in the left frontal cortex is associated with approach motivation, while right-dominant patterns often signal withdrawal or disengagement (Davidson, 1992). In neuromarketing research, tracking these asymmetries provides a real-time glimpse into how the brain engages positively or negatively to a stimulus—a powerful proxy for emotional receptivity and commitment. But what truly drives engagement at a primal level? Yet again, the answer is narrative. Stories are the brain's native language. They are efficient, memorable, and evolutionarily adaptive. A well-told story provides a structure that the brain instantly recognizes: a protagonist with a goal, a conflict, and a resolution. This framework aligns directly with our own life experiences, triggering profound cognitive and emotional resonance.

This simulation, often referred to as narrative transportation, is one of the most potent tools in the persuasion arsenal. When someone is deeply immersed in a story, they are less likely to counter-argue, more likely to mirror the emotions of the characters, and more open to the implicit messages being conveyed. Mirror neurons activate not just when we act, but when we observe someone else doing it, especially when that observation is embedded in a compelling narrative. Stories quite literally allow the audience to experience someone else's world as if it were their own.

There is a biological economy to storytelling. The brain is constantly bombarded with information, yet it prioritizes and retains data that is wrapped in emotion and causality—the hallmarks of compelling storytelling. In *f*MRI studies, stories that follow a clear arc with emotional highs and lows have been shown to synchronize neural activity across listeners (Hasson et al., 2008). This means that compelling storytelling can create a shared cognitive and emotional experience, aligning multiple brains toward a common interpretation. This is the neural definition of mass engagement.

Therefore, engagement through stories is also therapeutic. In fields such as narrative therapy and trauma recovery, storytelling is employed to help individuals reframe their experiences and restore coherence to their sense of identity. When we tell or hear stories, we not only process events cognitively but also resolve emotional conflict. In marketing, this translates to helping the audience envision a better future—one that the product or service helps unlock. Healing, in this context, is simply a return to narrative coherence.

But not all stories are created equally. For maximum engagement, stories must be relevant, emotionally textured, and marked.

While we have already discussed self-relevance as a core trigger, here it intersects with engagement by ensuring that the content resonates with the audience's self-schema. This mental framework defines personal identity. Emotional texture gives the story stakes and meaning, elevating it from mere information to lived experience. Marking refers to the psychological tagging of key moments, such as story surprises, turning points, or emotionally vivid scenes, that the brain bookmarks for later retrieval and recall. These moments create spikes in engagement and long-term memory encoding.

In a neuromarketing strategy, stories must be leveraged across touchpoints: product pages that introduce the brand's origin as a tale of struggle to success; testimonials that follow a before-and-after structure; and explainer videos that cast the customer as a protagonist overcoming challenges. These formats are neurosensory maps for engagement. Engagement itself is an ongoing process, built moment by moment. Each narrative beat serves as a micro-investment of attention, emotion, and relevance, moving the audience further along the motivational ladder.

Since the bottom-up progression begins in the primal brain, climbing through layers of cognitive commitment until the prefrontal cortex is the goal. By the time we reach the higher-order decision-making regions, we are already deeply immersed—psychologically hooked by the accumulated momentum of the story. The brain, reluctant to abandon sunk effort, stays loyal to the arc it has already begun to climb.

To illustrate the brain's resistance to changing course once a persuasive narrative has engaged its primal circuits and climbed to the prefrontal cortex, consider this analogy: Imagine pushing a heavy boulder up a hill. At first, it takes tremendous force to get it moving—just like activating the primal brain requires emotionally charged stimuli that bypass rational resistance. As the boulder gains momentum up the slope, you are not just pushing anymore; you are steering its trajectory. The limbic system begins to lock in emotional resonance, and the neocortex starts rationalizing the ascent.

Now, picture trying to veer that boulder sideways halfway up the hill, suddenly. It does not budge easily. That is because your energy, your focus, and your entire frame are committed to a path. Similarly, in the brain, this "cognitive inertia" stems from a combination of sunk psychological costs and narrative momentum. The brain resists new directions not because it lacks options, but because it has already invested in climbing this particular narrative slope.

This explains why, by the time persuasive content reaches the higher-order cortex, it is often too late to dislodge. The story arc is already entrenched, anchored by primal triggers and reinforced by emotional memory. Like a boulder in motion, our thoughts prefer consistency over redirection. This is the hidden gravity of the bottom-up effect.

Yes, engagement is cumulative. The more often a brand tells consistent, emotionally resonant stories, the more trust and affinity it builds. Over time, an engaging story shapes a "narrative identity," a personal storyline that the consumer internalizes to explain why the brand holds meaning in their life. This is the ultimate form of engagement: when the consumer's story and the brand's story begin to overlap, achieving synchrony between the brain circuits responsible for decisions, beliefs, and emotional commitments.

From a measurement perspective, combining frontal EEG asymmetry with biometric signals, such as heart rate variability, facial expressions, or pupil dilation, provides researchers with a multidimensional view of engagement. Peaks in engagement often correlate with moments of story resonance, narrative conflict, or emotional relief. These are the breadcrumbs that indicate when persuasion is occurring.

To summarize, engagement is a complex, measurable neurocognitive state rooted in motivation, emotion, and story. It is what happens when the brain says, "These matters. I am in." When marketers design content that activates this response—through meaningful narrative, emotional authenticity, and neural congruence, they do not just capture attention. They earn allegiance.

Before we explore our final trigger, it is essential to understand what truly sustains engagement once it has been sparked. Capturing attention is the opening move but sustaining it requires something more profound—a seamless cognitive experience that feels natural to the brain. This is where many persuasive efforts falter: they attract initially, but then drain our brain energy. The neural key to keeping attention—and converting it into preference—is not more stimulation, but more ease. With that, we turn to the final trigger in our neuro-persuasive arsenal: neuro-flow.

Trigger 12: Neuro-flow

If the brain had a user manual, "make it easy" would be on the first page. That is the essence of cognitive fluency: when information is easy to digest, it earns our trust. It feels right—even if it is not. Our primal circuitry is biased toward fluency, not out of laziness, but efficiency.

Neuro-flow, or what is often referred to as cognitive fluency, is not just about familiarity or even simplicity alone—these are separate and distinct triggers we discussed earlier. Instead, flow is the quality that permeates the entire experience of processing a message. It reflects how seamlessly information travels through our brain—from primal to rational—without generating resistance or cognitive strain. When messages are fluent—clear, simple, predictable—they require less energy to process. This aligns with our core survival programming, which is to preserve resources and minimize risks. Reber et al. (2004) found that fluent processing enhances not only comprehension but also likability, truth judgments, and aesthetic pleasure.

Neuro-flow is also more than simplicity. It is the absence of frictional state recognized in neuroscientific circles as cognitive flow (Csikszentmihalyi, 1990; Dietrich, 2004). This condition describes a seamless passage through which perception, cognition, and memory operate with minimal resistance, resulting in deep immersion and enhanced processing efficiency. In neuroscience, flow is associated with a unique state called "transient hypofrontality" (Dietrich, 2004), a temporary downregulation of the prefrontal cortex that enables more fluid and intuitive processing by subcortical regions, such as the basal ganglia and limbic system, both core structures of the primal brain. This shift frees the brain from the burdens of analytical oversight, allowing more instinctive, automatic processing to take over.

When we enter a flow state, we surrender control to the primal brain. It is a neurological efficiency upgrade: instead of slow, effortful decision-making, the brain relies on highly optimized neural circuits that have been developed through repetition and learning. Think about how we learn to ride a bike. At first, it is clunky, conscious, and mentally exhausting. Over time, the cerebellum helps the brain consolidate complex cognitive and emotional coordination into a kind of automated "mental muscle memory." Just as repeated physical practice leads to fluid movement, repeated exposure to fluent, well-structured messaging creates seamless processing pathways. The result? Experiences that feel easy, intuitive, and trustworthy qualities the brain not only prefers but instinctively seeks. The same applies to learning a language. What starts as memorization of grammar rules eventually becomes fluent, effortless conversation. Flow in behavior mirrors fluency in communication—and both depend on minimizing cognitive load.

In our neuromarketing lab, I have observed how subtle changes to a visual composition—its architecture, spacing, and harmony—can significantly impact engagement. The primal brain is visual, processing scenes first, before reading text. When a layout is cluttered, overloaded with competing shapes, excessive colors, or ambiguous symbols, it creates cognitive friction. Some objects are inherently more complex to recognize, while others compete for attention in ways that trigger distraction rather than direction. Visual confusion activates neural conflict monitoring systems, such as the anterior cingulate cortex, prompting the brain to pull away (Botvinick et al., 2004). But when design flows—when the number of elements is limited, the visual hierarchy is intuitive, and the placement leads the eye smoothly—the ventral striatum, the brain's reward center, lights up.

Effective design is neurologically efficient. It signals safety, predictability, and pleasure, reinforcing trust before a single word is even read. Think of flow as the neurological equivalent of excellent customer service. When a message is easy to understand, the brain feels served, not stressed. Moreover, even mild stress is the enemy of persuasion. The goal of neuro-flow is to remove interference. Even complex ideas can be delivered fluently when translated into brain-friendly formats, such as visuals, metaphors, stories, and chunked lists.

Significantly, neuro-flow enhances memory. Alter & Oppenheimer (2009) demonstrated that information processed fluently is not only better understood but also more likely to be remembered. This effect can be amplified through mnemonic techniques, methods specifically designed to make information more fluent and memorable. Rhymes, rhythms, acronyms, and visual imagery, though seemingly whimsical, produce measurable and fascinating effects on the brain. These mnemonic techniques enhance encoding and retrieval by capitalizing on the brain's affinity for patterns, sound, and imagery.

Functional imaging studies have verified that such methods engage not only the hippocampus but also activate multisensory cortical networks, enhancing both neural binding and recall speed (Yates, 1966; Maguire et al., 2003). This distributed processing not only strengthens memory but also makes recall more automatic and less effortful.

This is why rhythmic slogans stay with us like catchy tunes, why we instinctively recall jingles but forget corporate mission statements, and why well-designed infographics seem to explain themselves. These formats reduce cognitive friction and ride the brain's natural pathways for sound, pattern, and visual integration—turning fleeting messages into lasting memories.

From a neurobiological perspective, neuro-flow represents the optimal state for persuasion because it mirrors the brain's conditions during moments of deep immersion and effortless focus. The brain, driven by its hypothesized goal to streamline processing functions, constantly seeks to reduce the effort required for understanding and decision-making.

One striking confirmation of this hypothesis comes from the extensive study of Albert Einstein's brain. Contrary to widely held belief, Einstein's brain was not significantly larger, nor did it possess uniquely shaped anatomical regions. What set it apart, according to research published by Diamond et al. (1985) and further examined in Witelson et al. (1999), was the extraordinary level of myelination in his neural fibers—particularly in the corpus callosum and associative regions. The corpus callosum is the thick neural bridge connecting the brain's left and right hemispheres. In the context of neuro-flow, it plays a crucial role in integrating logical and emotional processing, allowing analytical structures and narrative elements to work in harmony.

Thus, when persuasion is fluid and immersive, this crosstalk between hemispheres becomes especially active, supported by associative regions that help synthesize meaning, emotion, and memory into a unified experience. Additionally, myelin—the fatty sheath that insulates axons—functions much like the rubber coating on electrical wires: it dramatically enhances the speed and efficiency of signal transmission. By reducing neural "leakage" and increasing conductivity, myelin ensures that information flows rapidly and precisely between brain regions, particularly across heavily trafficked pathways such as the corpus callosum. This insulation is essential for sustaining the fast, coordinated brain activity that underlies immersive states like neuro-flow. It facilitates superior integration between brain regions, supporting rapid, low-friction cognitive processing.

This scientific anecdote reinforces the idea that transmission efficiency—neural fluency—may well be a physiological cornerstone of achieving neuro-flow. The fewer barriers or bottlenecks within and between neural systems, the easier it becomes to reach and sustain states of immersion, insight, and fluid action. In this light, neuro-flow reflects the brain's deepest architectural preferences.

In fact, a growing body of research suggests that neural fluency may be more than just a brain feature; it could be a core component of intelligence itself. Studies show that individuals with higher cognitive ability often exhibit more efficient neural processing, using less energy to achieve more (Neubauer & Fink, 2009). Moreover, myelin plays a key role here—much like rubber coating enhances electrical conductivity, it accelerates and synchronizes neural signals.

In this light, neuro-flow reflects not only a psychological state but also a fundamental signature of optimized brain architecture. When a message aligns with well-worn neural pathways, it feels natural. It feels right because it mimics the conditions of flow. This brings us back to the twelve primal triggers.

By now, you probably understand that I consider neuro-flow a meta-trigger (a trigger above all triggers!), the gateway to persuasion. That makes each of the other eleven triggers a contributing factor to the conditions required for flow:

- **Self-Relevance:** Flow increases when the message feels personal. The medial prefrontal cortex activates when we perceive self-related content, facilitating faster processing.

- **Salience**: Our attention systems are wired to detect what stands out. Salience acts as a shortcut—it signals the brain to prioritize specific inputs, enhancing flow by eliminating distractions.

- **Change:** The brain is drawn to novel stimuli that signal shifts in the environment. Change captures attention and reorients processing resources, supporting flow by resetting cognitive focus.

- **Visual Priority:** The visual system processes images far more efficiently than text. Strategic visual elements enhance comprehension and minimize cognitive strain, therefore promoting visual flow.

- **First and Lasting Impressions**: Leveraging the primacy and recency effects ensures information is encoded and recalled with greater narrative flow.

- **Familiarity:** Recognizable patterns feel safe and predictable. Familiarity helps reinforce neural pathways, reducing resistance and enhancing receptivity.

- **Simplicity:** Simplicity clears cognitive clutter. When information is presented in a logical and visually clean manner, it promotes fluid understanding and a smooth flow.

- **Social Proof**: Seeing others' behavior reduces uncertainty. Social cues provide the brain with rapid decision shortcuts, enhancing processing flow.

- **Disruption:** Strategic breaks in pattern provoke re-engagement. Disruption can momentarily increase cognitive load, but it can also sharpen attention and refresh the flow if managed well.

- **Anticipation**: The brain is a prediction engine. When messages satisfy expectations, dopamine is released, supporting the flow of reward and retention.

- **Engagement:** Active mental participation sustains immersion. Whether solving, predicting, or co-creating meaning, engagement fuels cognitive momentum and fosters a state of flow.

Chapter Summary

To summarize, Chapter 4 introduces neuro-flow as the apex state of persuasion. This meta-trigger emerges only when all the primal elements align to create a message that feels effortless, immersive, and emotionally congruent. Neuro-flow is not a standalone effect; it is the compounded result of activating multiple primal systems in synchrony, allowing the brain to shift from decoding to absorbing, from skepticism to surrender.

In total, twelve primal triggers play a vital role in this orchestration. Self-relevance ensures the message feels personally significant. Salience directs attention to what stands out in a noisy world. Change creates the neural friction needed to perceive difference and make choices. Visual imagery engages the brain's dominant sensory processor, enhancing comprehension. First and last impressions shape how experiences are encoded and recalled. Familiarity lowers resistance by reducing perceived threat. Simplicity eases cognitive load, making the message easier to integrate. Social proof validates a message through perceived consensus and the safety that comes with numbers. Disruption resets attention by breaking expectations and forcing reappraisal. Anticipation builds a sense of forward momentum, drawing the brain into what is next. Moreover, engagement, far from being a vague feeling, is a measurable neural state of readiness and emotional investment—it signals the brain's decision to care.

When these triggers work in concert, they create a cognitive and emotional glide path. The message does not feel like work to process—it feels like something the brain is already inclined to follow. That is the essence of neuro-flow. It arises when the architecture of the message mirrors the architecture of how the brain prefers to operate: rhythmically, emotionally, visually, and with minimal resistance.

Neuro-flow is more than high engagement—it is a state of neural alignment. It allows a message to move swiftly through attentional, emotional, and decision-making systems without a bottleneck or breakdown. In this state, persuasion does not feel like effort; it feels like resonance. The brain leans in, not out.

In a world saturated with clever distractions and cognitive overload, Chapter 4 reveals a counterintuitive truth: the most persuasive messages are not the most complex; they are the most synchronized. Neuro-flow is the signal that all systems are go. It is persuasion in its most elegant form—an experience that does not just capture the brain but carries it. Now that we have explored how neuro-flow emerges from the orchestration of primal triggers, the next step is learning how to design messages that consistently activate them. This is where structure meets strategy. In the following chapter, we introduce NeuroMap®—a science-based framework developed to map the journey of persuasion through the brain systematically. If neuro-flow is an experience of seamless influence, NeuroMap® is the blueprint that makes it repeatable. Let us turn to the model that has helped thousands of brands turn neural principles into persuasive precision.

A Quick Primal Brain Exercise

Grab a piece of persuasive content—a web page, sales deck, email, or ad—and run it through this rapid neuro-checkpoint. Ask yourself:

Self-Relevance: → Does the message speak directly to the needs, fears, or identity of the target audience?

Salience: → Does it stand out instantly? Is there a bold headline, striking claim, or visual hook?

Change: → Is there a clear "before and after" or contrast that makes the choice or transformation obvious?

Visual Priority: → Can the brain "see" the message? Are ideas communicated with strong visuals or sensory language?

First and Lasting Impressions: → Does it open and close with memorable, emotionally resonant moments?

Familiarity: → Does the message reference known concepts, symbols, or patterns that make it feel safe or recognizable?

Simplicity: → Is the message stripped down to its core? Can someone get the point in under 8 seconds?

Social Proof: → Does it show that others have trusted, used, or benefited from it?

Disruption: → Is there a surprise element that breaks expectation and forces re-engagement?

Anticipation: → Does the message create curiosity or a sense of "what's next"?

Engagement: → Does it create emotional or mental investment? Are you pulled in enough to want more?

Neuro-Flow: → Taken together, does the message feel smooth, immersive, and rewarding to process?

Final Thought: Choreography of instincts

When the primal triggers align, they not only deliver information but also create momentum. Neuro-flow is the brain's signal that everything is working as it should: attention is captured, emotion is engaged, resistance is quieted. In this state, the message does not feel like something to evaluate—it feels like something to follow. And in a distracted world, that feeling is persuasion's rarest and most powerful currency.

CHAPTER 5: Mapping Persuasion with NeuroMap®

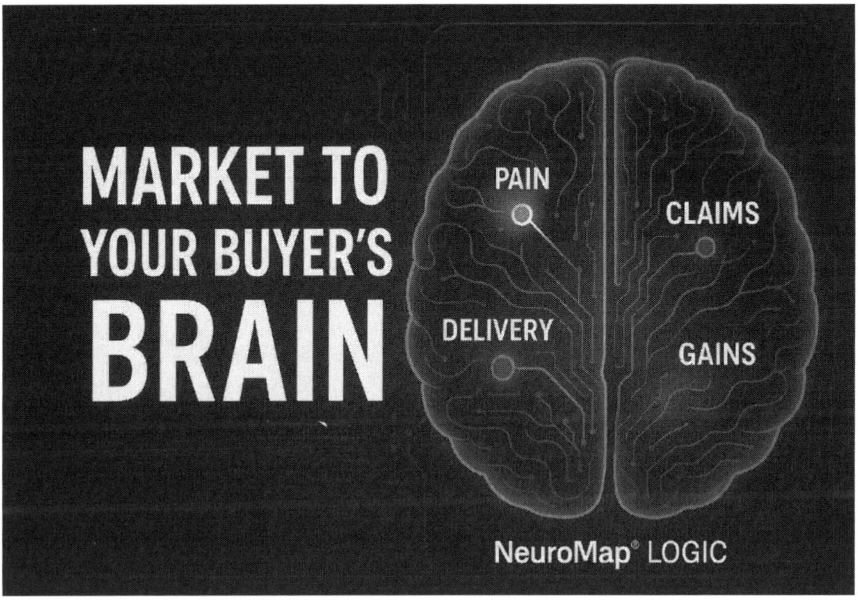

If I cannot build it, I do not understand it.

Richard Feynman, Ph.D.
Nobel Prize-winning physicist and science communicator

In the early days of neuromarketing, the landscape was filled with promise but littered with confusion. Marketers had long known that traditional approaches—rooted in demographic segmentation, rational messaging, and gut feeling—were failing to deliver predictable, replicable results. They understood the "what" of consumer behavior but not the "why."

Meanwhile, the field of neuroscience was blossoming with innovative technologies—*f*MRI, EEG, and biometrics—that could provide real-time insights into the brain. However, bridging the gap between brain data and business decision-making was akin to trying to navigate the Amazon with a tourist map of Paris. Researchers could detect attention and emotional arousal, but they could not tell a CMO which headline would convert better or why the same ad failed miserably in one market and thrived in another.

What was missing was a model—a usable, teachable, actionable framework that connected the dots between brain function and buyer behavior. A map not just of geography, but of *meaning*. One that did not merely describe persuasion but predicted it.

This was not just scientific curiosity or wishful thinking. It was a practical necessity. Every year, billions were spent on advertising that produced little more than noise. Creative teams chased originality without considering biological plausibility. Strategists spun narratives without grounding them in the way people *actually* make decisions using their brains. In this chaos, Patrick and I saw the need for a structured model—not just to decode persuasion, but to engineer it.

That is where the journey of NeuroMap® began.

The Genesis of NeuroMap®

Patrick and I were driven by both frustration and fascination. We had witnessed too many failed marketing campaigns, too many hollow pitches that dazzled creative directors but left audiences unmoved. We wanted to stop guessing and start grounding persuasion in the science of how the brain works. We both recognized that marketing was not about crafting more clever messages; instead, it was about learning to speak the brain's native language.

The journey, however, was anything but smooth. Patrick and I often clashed on how best to distill the complex science into a practical framework. Debates would escalate, sometimes ending only after an intense match on the tennis court—our unspoken truce arena. Yet despite our disagreements, we shared a relentless determination to build a persuasion model that could stand the test of time and scrutiny. NeuroMap® was the product of both our collaboration and our creative friction.

With our combined backgrounds in sales, marketing, consumer research, computer science, psychology, and neuroscience, we set out to build a model that would make persuasion tangible. It needed to be easy to teach, rigorously testable, and effective. We immersed ourselves in the literature of cognitive science, evolutionary biology, and behavioral economics. That work culminated in the birth of NeuroMap®: a step-by-step framework designed to help communicators align their message with the brain's hardwired decision-making pathways.

We conducted field research, ran experiments, and pressure-tested our ideas in client meetings, workshops, and thousands of training sessions. Yes, together, we have personally delivered well over 4000 training sessions in the last 23 years. And counting!

But NeuroMap® was never meant to be an endpoint. Even in those early days, Patrick and I realized that it was only the first attempt at translating the messy, beautiful, and unpredictable human behavior into a model that could guide persuasive communication for decades.

What Worked: Clarity, Structure, Replicability

From the model's inception, NeuroMap® 's most significant contribution was its structure. It was meant to offer marketers, sales professionals, and communicators a way to decode the decision-making process of the brain and translate that into a step-by-step sequence: first, before diving into the process of persuasive messages, NeuroMap® insists that you must "Diagnose the pain" you are on a mission to pulverize with your products, services, or ideas. Second, you must "Differentiate your claims", unpacking up to three credible benefits that hold unique value. Third, your claims must "Demonstrate the gain", compelling evidence that proves you can deliver what you promise.

These three pillars were not just catchy (notice the rhyme for PAIN/CLAIM/GAIN that took weeks to settle on). They also mapped directly onto the primal brain's obsession with avoiding threats, making quick decisions, and assessing value. The simplicity of the model made it sticky. The alignment with neuroscience made it credible. The framework itself made it teachable, repeatable, and adaptable across sectors—from pharmaceuticals to finance to education, and including collaborations with companies such as GE, Intel, PayPal, eBay, Microsoft, Siemens, Avon, Esurance, and many other Fortune 1000 brands worldwide.

We heard it again and again from our clients: "This finally makes sense." Even better, people did not need to be neuroscientists to understand or use NeuroMap®. They needed a roadmap that mirrored how real brains—outside of theoretical textbooks—behave in the marketplace.

Meanwhile, even as our persuasion model gained traction, Patrick and I understood its limitations, shared by all behavioral models. The brain does not process persuasion as a tidy sequence of steps. It operates as a living network, with countless signals, memories, emotions, and biases colliding in real time. A model like NeuroMap® can chart the most likely path, but it cannot predict every turn or execute the journey for you. That requires the marketer's hands-on work: asking challenging questions, reading the audience, testing relentlessly, and adapting when the unexpected happens. Therefore, we taught our customers to anticipate and adapt the model to the reality of their own situation. Flexibility was key.

Let me be honest, we also encountered significant resistance to the model, particularly from advertising executives and traditional marketers who were heavily invested in older, more familiar paradigms. Many had built their careers on intuition and creative flair rather than science. To them, adopting a neuroscience-based model felt not only threatening but almost heretical. Clinging to archaic frameworks seemed safer than embracing a disruptive, data-backed approach. After all, if you admit your messaging is not aligned with how the brain works, what does that say about the millions spent on campaigns built on shaky assumptions?

Patrick and I had to navigate skepticism not just from clients, but from entire industries that were reluctant to question their sacred cows. Yet it was this very pushback that reinforced our commitment to fine-tuning a model that did not just sound good in a pitch deck—it had to hold up under scrutiny, deliver results, and evolve. Which is why we had to ask: What happens when persuasion must operate in real time—

dynamic, adaptive, and responsive to context? What if the message must evolve not just across markets, but in milliseconds, as attention shifts and emotions fluctuate?

Beyond the Original Map: Why We Needed AI

That is when I began dreaming of an assistant—not just human but powered by AI—that could run the NeuroMap® engine in its sleep and eventually outthink us both. Despite its elegance, we realized the model could not fully capture the complexity of human attention, emotion, and decision-making in the digital age without further evolution and technological advancements.

NeuroMap® began as a four-step framework, then later expanded to integrate six primal stimuli, six persuasion elements, and seven persuasion catalysts. But it was poised for another significant leap.

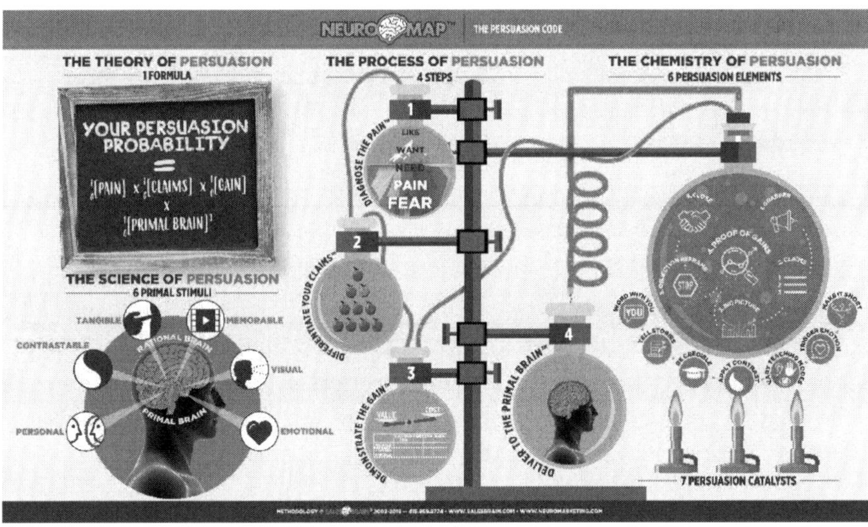

Figure 22: NeuroMap® (SalesBrain)

Two seismic shifts accelerated this transition. First, the COVID-19 pandemic hit, halting lab-based neuromarketing studies for two years. This forced the industry to rethink data capture strategies, shifting from in-person measurements to online data collection and then into predictive modeling. Second, this pivot accelerated investment in AI and transformed the field forever. In that context, building an additional data-centric tool, such as the NeuroQuadrant™ Analysis, made perfect sense—especially as the production and scoring of campaigns became faster, more efficient, and more precise than ever before.

Meanwhile, my interest in artificial intelligence did not appear overnight. It began during my earliest neuromarketing studies, when I realized the potential richness and overwhelming nature of EEG, eye tracking, galvanic skin response, and other biometric data. We were dealing with noisy, multidimensional datasets—fluctuations in voltage, heat maps of attention, and microexpressions captured in milliseconds. Extracting meaning from this deluge was like listening to a whisper in a thunderstorm.

And yet, those whispers were everything. They revealed why a headline failed, why an ad provoked irritation instead of curiosity, and why one product display arrested attention while another faded. Unlocking them required a tool that would be smarter, faster, and more systematic than any human analyst.

I realized the future of persuasion would not just rely on theory—it would depend on computational power. Neuromarketing research is not easy, and without powerful software to analyze and interpret results, many insights would remain hidden. Our audiences were no longer passive; they were navigating dynamic, multi-screen environments where messages had seconds to succeed. This demanded adaptability, nuance, and speed—a tireless interpreter of brain cues unclouded by guesswork.

AI offered this potential. Not as a silver bullet, but as a relentless co-pilot that could ingest neuroscience patterns, decode behavioral signals at scale, and personalize persuasion strategies to each mind. The goal was not to replace NeuroMap®—it was to elevate it, to give it memory, agility, and predictive power.

The seed for this leap was planted in the lab, as we began scoring campaigns and processing neurodata through AI. By quantifying primal responses to real messages, we could benchmark them, measuring the "distance" between two messages—not just which was stronger, but why. This fusion of scoring, neuroscience, and machine learning led to the creation of a tool designed to map motivational mindsets with unprecedented clarity.

In the world of persuasion, clarity is a powerful tool. That is why, in addition to models like NeuroMap® and dual-system thinking, this chapter discusses another essential compass: the NeuroQuadrant™

System. This framework maps how consumers interpret messages along two axes, primal impact on the vertical and rational impact on the horizontal. It scores the strength of bottom-up, instinctive engagement against top-down, cognitive evaluation. This enables marketers to predict not only how a message will be received but also the type of persuasive state it will trigger.

NeuroQuadrant™: Understanding Your Persuasion Zone

The "Persuasion Zone" is a field of influence that maps messages according to their ability to activate both the primal and rational brain

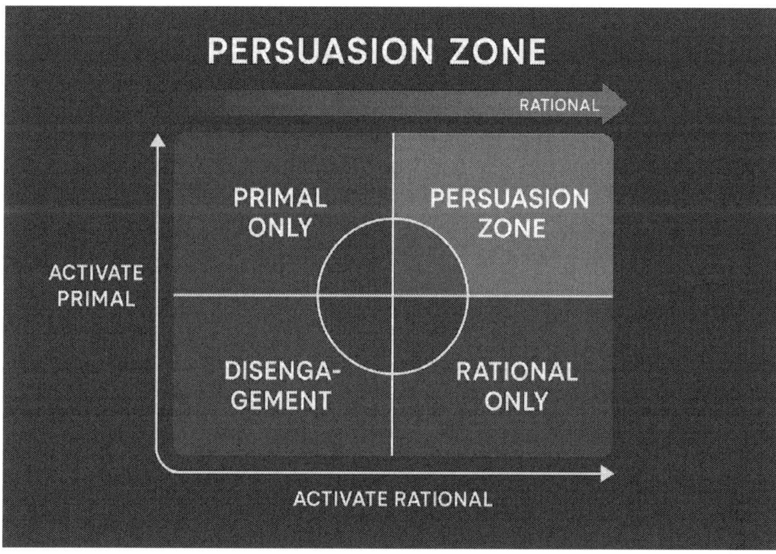

Figure 23: The Persuasion Zone

systems. The 4 NeuroQuadrants™ reflect how a message engages—or fails to engage—our dual decision-making architecture.

NeuroQuadrants™ I: The Persuasion Zone

This quadrant indicates a perfect alignment between instinctive Primal Brain activation and logical Rational Brain appeal. This combination produces maximum persuasive power, high retention, and decisive action. Consider a home security ad that opens with a brief clip of a door left ajar at night, then immediately shows a simple plan, a price, and testimonials from neighbors. The primal system flags the scenario as personally relevant and risky, while the rational system receives clear evidence and a credible offer. The result is both urgency and justification.

Recommendation: Optimize this zone by maximizing emotional contrast and rational clarity. For example, pair a visceral scene with a one-line benefit and a price that is easy to compute.

NeuroQuadrants™ II: Rational Only

Messages here are intellectually sound but emotionally flat. Without primary engagement, the bottom-up effect is weak, which drives high cognitive effort, confusion, and poor memorability. A typical example is a B2B software page that leads with a dense feature grid and four charts yet shows no human stakes and no visual story. Readers may concede that the product appears capable, but the message is not felt and therefore not retained.

Recommendation: Raise primal appeal by leveraging the six stimuli. Make the message personal by naming the pain, make it visual with a three-frame storyboard, and make it emotional with a brief consequence statement. Only then should detailed proof points appear.

NeuroQuadrants™ III: Dis-engagement

This is a total persuasive failure. The message lacks both emotional resonance and logical justification, so audiences neither feel nor understand the value. Picture a generic banner that reads 'Innovative Solutions for Tomorrow' over a stock skyline. There is no pain, no proof, and no path to gain.

Recommendation: rebuild from scratch using the six stimuli to establish both primal and rational relevance. Start with a specific audience, reveal a concrete loss that matters to them, introduce a sharp contrast with the current alternative, and show a tangible demonstration that creates a small win in seconds.

NeuroQuadrants™ IV: Primal Only

These messages are emotionally captivating but logically incomplete. A strong primal impact grabs attention, yet without rational

support, persuasion remains incomplete. A charity video that makes viewers tear up but never states what one donation accomplishes is a typical case. Audiences feel compelled but hesitate because the cognitive story is missing.
Recommendation: strengthen narrative structure, sharpen contrast, and provide a simple plan that turns emotion into action. Show the first milestone a gift unlocks, for instance, one kit delivered or one hour of tutoring secured.

The 3 Stages of the Bottom-up Effect

According to NeuroMap®, a message rarely lands in the

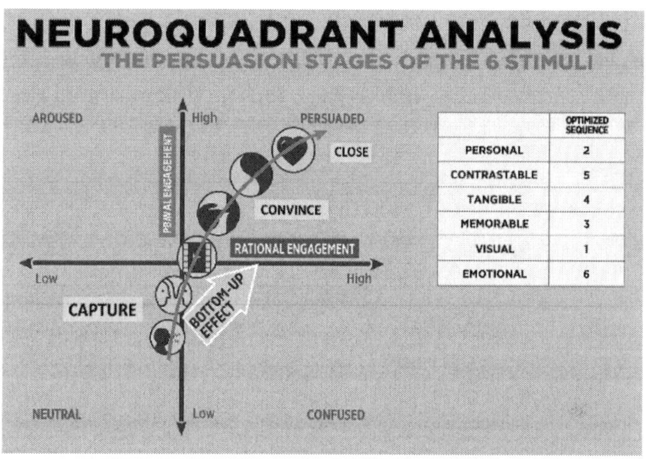

Figure 24: NeuroQuadrants™ Analysis

"persuaded" zone by accident. It progresses through a path that begins with instinctive capture and ends with a confident decision. Effective persuasion unfolds in three stages: Specific stimuli best serve to *capture, convince, and close,* and each stage. The one overarching goal remains constant: to reach the northeast corner of the map where persuasion is most potent.

Capture

Capture is driven by primal triggers that seize attention in milliseconds, often before the audience consciously registers the content.

Personal and visual elements contribute the most to this initial engagement. A health app aimed at runners over forty could open with a tight shot of knees hitting the pavement, accompanied by the line: 'For runners 40-plus who want pain-free mileage.' A credit card teaser that flashes two contrasting images — a trip booked, and a trip cancelled — invites immediate self-reference and visual contrast. In both cases, the viewer's attention is not requested; it is compelled. Functional diagnostics at this stage include eye tracking for first fixation and dwell time, as well as GSR for arousal peaks. If these signals are weak, the message is not yet persuasive, regardless of how solid the logic may be.

Convince

Convince builds on the captured attention by encoding belief through memorable and tangible appeals. Here, the task is to move from "I noticed this" to "I believe this belongs to me." Memorable content employs simple framing devices and sticky analogies, for example, a cybersecurity firm likening multi-factor authentication to a double lock on a front door. Tangible content turns abstraction into evidence, such as a 20-second product demo that shows a spreadsheet filling itself or a bar chart comparing error rates before and after adoption. In retail, a makeup brand can show a split-face photo and a one-line explanation that names the ingredient responsible. EEG and recall tests are helpful in this context because they assess cognitive integration and later memory, both of which predict whether the belief will be defended in conversation.

Close

Close delivers the final push into decision-making, where contrastable framing and strong emotions work together to commit the audience to act. Contrast clarifies the trade-off; for example, a before-and-after calculator that shows the money or time gained, or a two-choice layout that clearly labels the inferior option. Emotion provides the nudge that resolves hesitation, such as a relief image that appears once the viewer selects a plan, or a brief story of a customer who overcame the exact obstacle the buyer now faces. Calls to action should be precise and frictionless; think 'Book the 15-minute fit check' rather than 'Learn more.' Behavioral signals at this stage include clicks, form starts, and micro-conversions. Meanwhile, facial coding can help verify whether the closing emotion is relief, pride, or hope, rather than anxiety.

How the Six Stimuli Affect the 3 Stages

Through hundreds of campaigns, the same pattern recurs. Personal and visual cues dominate the *capture stage* because they are processed rapidly by the primal gatekeeper and act like a pass to the rest of the message. Memorable and tangible cues do the heavy lifting in the *convince stage*, because they encode the claim into a story the brain can store and retrieve. Finally, contrastable framing and emotions drive the *close stage because* they narrow choices and apply the energy needed to cross the final gap. Here are examples of how NeuroMap® can be used across different sectors:

Consumer packaged goods.

A beverage launch features a six-second pre-roll that shows a hand twisting a cap, and the word "Calm" appears as the fizz settles. The landing page opens with three icons: "Less Sugar," "Added Magnesium," and a recyclable bottle, followed by a taste test chart that compares the sweetness of the product to that of its competitors. A limited trial pack and a timer are included for free shipping.

Financial services.

A retirement platform opens with a calculator that fills in a projected shortfall using average savings for people of the viewer's age. The page then displays two side-by-side stories: one path that maintains current contributions and one that adds 3 percent. A video testimonial from someone with a similar profile provides emotional social proof. The close offers a prescheduled call slot and a guarantee to create a plan in 30 minutes. The progression pulls visitors by adding personal, visual, and emotional fuel to a previously rational argument.

Healthcare.

A clinic uses geotargeted ads that state "Same-Day Relief for Migraines" within a five-mile radius. The ad links to a page featuring a three-step diagram, intake information, treatment details, follow-up information, and the associated costs. A short patient story concludes with the line, 'I felt like myself again in two hours.' Appointments are booked directly from the page. This sequence moves prospects from Capture to Close by delivering personal relevance first, then tangible proof, then emotional resolution.

B2B software.

A workflow tool displays a 10-second screen capture showing how a chaotic inbox is transformed into a clean Kanban with one click. The headline reads: 'Turn Buried Requests into Finished Work.' Below it, a two-column layout lists the current alternative on the left and the tool on the right with three measurable differences. A customer quote provides the final emotional lift. Trials spike because the message now clears both the primal and rational bars.

Moving Messages Across NeuroQuadrants™

Different audiences and contexts place people in different zones. A new parent may sit in NeuroQuadrants™ IV when thinking about their

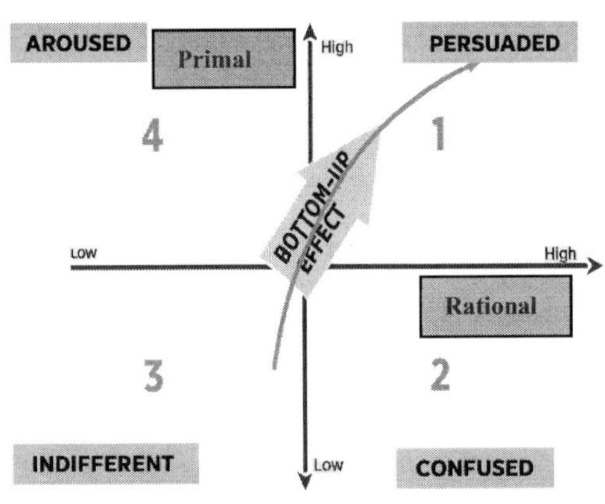

Figure 25: Persuasion Journey

child's safety, but a college student may dwell in NeuroQuadrantsTM II when planning for college. A loyal iPhone buyer may feel indifferent to NeuroQuadrants™ III when seeing ads for Android phones, but a new car buyer can be propelled to NeuroQuadrants™ I after test-driving a BMW. The map does not replace segmentation. It enriches it by adding a motivational lens. Campaigns can also move deliberately across zones: dramatize an urgent problem to capture (I), offer a preventative plan to convince (IV), and finish with a vision of long-term success to close (IV to I). Messages that try to straddle zones without clarity often fail because the primal brain cannot resolve mixed signals. Select a primary zone for each asset and maintain consistent cues.

By plotting messages on the NeuroQuadrants™, diagnosing their stage along the bottom-up path, and concentrating on the right stimuli at the right time, marketers can guide a clean progression from attention to conviction to commitment. The destination is singular: persuade on this map, which means high Primal engagement and high Rational engagement working together to produce choices that feel right and make sense.

Teaching Machines to See Persuasion

Meanwhile, many years before generative AI dominated headlines, our lab had adapted computer vision to study how people perceive persuasive media. Historically, computer vision has been one of the most influential forces shaping artificial intelligence. Its roots go back to the 1960s, when early AI pioneers dreamed of teaching machines to interpret the visual world—an ability that humans perform so effortlessly that we often overlook its computational complexity. In those days, researchers tackled simple problems, such as recognizing shapes or detecting edges in images. By the 1980s and 1990s, advances in mathematical modeling, image segmentation, and pattern recognition enabled computer vision to automate real-world tasks in manufacturing, medicine, and robotics.

The arrival of machine learning—and the development of intense neural networks in the 2010s—marked a turning point. Suddenly, instead of hard-coding vision rules, systems could "learn" from massive datasets to recognize faces, track movements, and interpret complex scenes. This made computer vision a foundational discipline in AI, influencing fields from self-driving cars to biometric authentication.

Our lab leveraged this capability in a way that had little to do with making a creative robot. Instead, we adapted computer vision as a neuromarketing tool to study how people look at persuasive media. By using eye-tracking, gaze mapping, and emotion-recognition algorithms, we could analyze viewers' unconscious attention patterns—measuring not just whether they looked, but what caught their gaze and for how long. These insights helped us connect visual attention with the brain's decision systems, particularly the primal brain's fast, emotional responses.

We aimed to build a persuasion evaluator that could read visual features the way a brain does and forecast where a message is likely to land on the NeuroQuadrant™ system. We trained models to score elements such as contrast, motion, face presence and expression, gaze direction, symmetry, text density, and object prominence. Those scores were then translated into two estimates that matter most for persuasion: expected Primal pull and expected Rational friction. In practice, this resulted in a rapid pretest. A creative asset could be flagged as likely to belong to NeuroQuadrants™ II, IV, or I before investing in distribution or full neurometrics.

However, this approach only works when the model is anchored to a cognitive compass. The rubric tied each feature back to the six stimuli and to the two axes of the NeuroQuadrants™. For example, strong visual contrast and a human face with visible affect often lift the Primal estimate. Clean layout, low reading load, and a concrete claim lift the Rational lift. The result is not a verdict on creativity. It is an early signal that guides iteration before costly testing.

Benefits include speed, scale, and comparability across versions and markets. Limits include context sensitivity, dataset bias, and the fact that computer vision sees pixels, not purpose. That is why AI scoring complements but never entirely replaces lab measurement. To transition from pattern to proof, we still require ground truth from both the body and the brain.

To validate and refine machine estimates, we rely on methods that capture the biological signatures of attention, emotion, and cognition. The next chapter explains how and when to use them, and what each tells you about the three stages of persuasion: Capture, Convince, and Close.

In Chapter 6, you will learn:

- **Autonomic measures for Primal signals.** Galvanic skin response and heart metrics reveal arousal patterns. Respiratory sinus arrhythmia and heart rate variability add nuance to the understanding of regulation.

• **Ocular and facial measures to see where and how people look**. Eye tracking reveals first fixation, dwell time, and scan paths. Facial coding reads valence shifts and discrete expressions at the frame level.

• **Voice and endocrinal markers for stress and engagement when appropriate.** Layered voice analysis can detect tension. Hormonal assays can index affect in specialized studies.

• **Cortical measures for Rational processing**. EEG provides millisecond timing of attention and workload. MEG and *f*MRI provide detailed maps of cortical involvement when spatial precision is crucial.

• **Implicit cognition for attitude strength**. The Implicit Association Test and related tasks reveal associations that people may not be aware of or are unwilling to report.

We will compare methods on cost, ecological validity, temporal and spatial resolution, and ethical requirements. Most importantly, we will show how to combine them. In many projects, a light stack is enough. Eye tracking and GSR can confirm *Capture*. Short recall and comprehension tests can confirm *Convince*. A simple conversion task, paired with facial coding, can confirm Closure. In complex work, EEG or *f*MRI adds depth and helps isolate the neural correlations of the two axes.

Chapter Summary

Chapter 5 explains how AI became a turning point in Neuromarketing, as it transitions from theory to an interpretive framework. Persuasion becomes not only a psychological phenomenon but a mappable process that is systematic, testable, and scalable. The centerpiece of this shift is the NeuroQuadrant™ System, a neuroscience-grounded framework that aligns persuasive messaging with the way the brain decides. Developed through decades of research and application, this additional tool gives marketers, designers, and communicators a structured path to move a message from exposure to action.

The model is built on the four essential steps that mirror the brain's sequence of response. First, seek relevance to a pain, the felt need, fear, frustration, or problem that commands attention. Second, identify claims so the message stands out in a noisy field. Third, demonstrate the gain so the brain can see and feel a tangible, emotionally resonant benefit. Fourth, deliver to the Primal Brain through simple, visual, emotional, and immediate encoding. The chapter emphasizes that many messages fail not because the logic is weak, but because they are presented at the wrong cognitive level. Rational appeals that skip the primary gatekeeper get filtered out as irrelevant. NeuroMap® ensures that messages first resonate at the instinctive level and then progress to higher-order reasoning.

The NeuroQuadrants™ System informs us how to structure a persuasive message and identify where the audience is likely to receive it. The map plots messages along two axes: Primal impact on the vertical axis and Rational impact on the horizontal axis. The result is four persuasion zones that describe common states: high Primal and high Rational, where people are persuaded; high Rational but low Primal, where logic is clear but cold; low Primal, where indifference prevails; and high Primal but low Rational, where emotion is strong but unfinished. Knowing a message's position allows teams to choose the next best move, whether that is to add personal and visual capture, encode belief with memorable and tangible proof, or close with contrast and emotion.

Together, NeuroMap® and the NeuroQuadrant™ System form a dual-lens system for persuasion, one for structure and one for calibration. Used ethically, this integrated approach replaces guesswork with grounded neuroscience, converting attention into action.

A Quick Primal Exercise for You

Before delivering a critical message, pause and verify the sequence. Have you diagnosed the precise pain your audience feels? Have you framed a credible claim that creates a clear contrast between the current pain and a better future? Can you demonstrate the gain with evidence, a story, or a visualization? Finally, what tools or data can help you validate that your message aligns with how the brain responds? Could an AI assistant help you simulate or test the neurology of your message?

Final Thought

Persuasion is not fading, but the methods we use to decode and deliver it must evolve. The fusion of brain science and machine intelligence offers not only efficiency but empathy on a scale if we wield it wisely. NeuroMap® is a compass. Neuromarketing AI is fuel. The next chapter explains the measurement toolbox that anchors this system in biology and behavior, allowing pixels and words to be tied to signals and choices with rigor.

CHAPTER 6: Measuring the Mind Activity

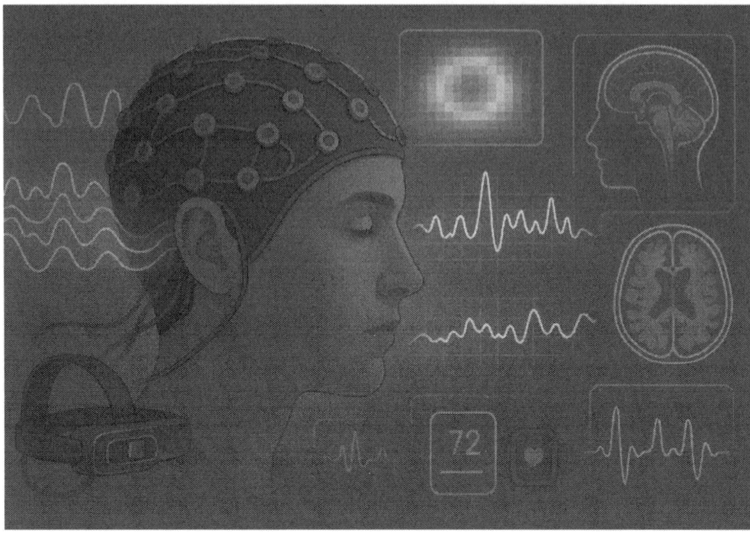

The initiation of action occurs in the brain before there is any awareness of wanting to act.

Benjamin Libet, Ph.D.
Physiologist

For centuries, persuasion was more art than science. Marketers and communicators relied on instincts, focus groups, and post-campaign metrics to gauge whether their message had landed. However, these tools barely scratched the surface. They offered delayed, diluted signals—like trying to judge a rollercoaster's thrill from a blurry photo taken after the ride ended.

Then came neuromarketing, with its promise of opening the black box of the brain. For the first time, we were able to observe the mind respond in real-time. Emotions, attention, and cognitive overload—these constructs and outcomes were no longer just abstract ideas. They became data. The instruments of this revolution? Facial coding, GSR, eye tracking, EEG, functional magnetic resonance imaging (*f*MRI), and a growing toolkit of neurophysiological measures offer a different lens through which to observe the dance of neural activity triggered by persuasion.

If you want to influence behavior truly—not just track it after the fact—you cannot afford to stay at the surface. Neuromarketing methods pull back the curtain on the brain's decision engine, revealing not just what people say or do, but why they do it—before they even realize it themselves. But not all tools speak the same neural language. Some tap directly into the primal circuits that trigger instant action, while others trace the rational deliberations that follow. Understanding how each method works and which part of the brain it illuminates is the difference between guessing and knowing. So let us get under the hood.

Primal Brain Methods

Before diving into specific tools like GSR or eye tracking, it is essential to grasp the physiological terrain they explore. *Primal Brain Methods* refer to measurement techniques that capture subconscious, instinctive responses governed by the **Autonomic Nervous System (ANS),** the fast-acting neural infrastructure wired for survival, not reflection.

The ANS splits into two main branches: the **sympathetic nervous system**, responsible for arousal and action ("fight or flight"), and the **parasympathetic nervous system**, which promotes recovery and calm ("rest and digest"). Together, they orchestrate a hidden ballet of signals in the body—fluctuations in heart rate, skin conductance, pupil dilation, and facial micro-movements—that tell us how a person feels before they can articulate it.

These methods do not rely on self-report, memory, or reasoning. Instead, they intercept the body's raw output as it responds to stimuli because the Primal Brain does not speak in words—it speaks in reactions.

Using these tools, we can ask: Did the ad grab attention? Did the customer feel stress, excitement, or boredom? When exactly did those reactions occur? Some core advantages of Primal Brain methods include:

- **Unconscious validity**: They reveal truths participants cannot fake or forget.
- **Temporal precision**: With millisecond-level tracking, we can pinpoint *when* a message succeeds or fails.
- **Cross-cultural robustness**: Basic ANS responses are evolutionarily conserved—what triggers arousal in Tokyo, triggers it in Toronto too.

In this chapter, we will explore the most validated ANS-driven techniques, beginning with endocrine studies, which measure not just what we think, but also what we feel in our bloodstream. However, keep in mind that each method offers a different perspective. The magic happens not in choosing one, but in triangulating the truth across multiple bio-signals. That is how you decode the primal mind.

Endocrinal studies: when blood carries our emotions

As a scientist devoted to understanding persuasion, I have long believed that what people *say* about ads is far less revealing than what their brains—and indeed, their bodies—*do* in response. While tools like EEG and facial coding illuminate cognitive and expressive patterns, they only tell part of the story. The biochemical layer—specifically the hormonal response—is where persuasion becomes visceral.

In collaboration with Dr. Paul Zak, a pioneer in neuroeconomics, I explored this dimension by examining the role of oxytocin, a neuropeptide widely recognized for its role in facilitating trust and social bonding. Our hypothesis was clear: emotionally resonant messages could trigger the release of oxytocin, thus biologically predisposing viewers toward empathetic and prosocial behaviors, including donation intent.

To test this, we designed a study that required participants to view emotionally charged public service announcements (PSAs) while undergoing both biometric monitoring and blood sampling to assess oxytocin levels. This was no minor undertaking. Because of the invasive nature of the study, two Institutional Review Boards approved one at Fielding Graduate University and one associated with Dr. Zak's lab at Claremont Graduate University—to meet stringent ethical guidelines. This research was a principal component of my doctoral dissertation, *The Neurophysiological Effect of Emotional Ads on the Brains of Late Adolescents and Young Adults* (Morin, 2014).

The results were illuminating. We found that PSAs designed with strong narrative structure and emotional valence significantly elevated oxytocin levels in participants. Importantly, these increases correlated with both self-reported empathy and a statistically significant rise in actual donation behavior. In short, hormonal data helped predict persuasion outcomes with a level of granularity and reliability that traditional self-report methods cannot match.

Mechanistically, oxytocin modulates activity in the limbic system, particularly the amygdala and hypothalamus, attenuating fear responses while amplifying affiliative behavior. This hormonal shift effectively lowers cognitive defenses and facilitates emotional openness—an ideal state for persuasive messaging.

Yet I must emphasize: this type of endocrinal research is too invasive for commercial use. Drawing blood is a clinical procedure, not a marketing tactic. While the insights are powerful, they are best reserved for academic inquiry under strict ethical safeguards.

Still, these findings have enduring value. They affirm that persuasion is not just a cognitive process; it is profoundly biochemical. The study showed that trust, empathy, and decision-making can be traced not just to cortical activity but to the bloodstream itself. It provided hard evidence that emotion-driven storytelling does not just "move" us metaphorically—it alters us chemically.

This line of inquiry forms a foundational layer of what I now call Neuromarketing AI: an integrative science where primal reactions, emotional resonance, and physiological data converge to reveal how we are wired to say "yes."

Galvanic Skin Response (GSR): Sweating the Truth

Galvanic Skin Response (GSR), also known as Electrodermal Activity (EDA), offers a microscopic window into the silent symphony of emotional arousal. Originating from early polygraph use, GSR measures minute fluctuations in skin conductance caused by the activation of eccrine sweat glands, innervated by the sympathetic nervous system.

What makes GSR scientifically potent is its bifurcation into **tonic** and **phasic** signals. The tonic component represents the baseline level of skin conductance over time—think of it as the emotional climate of a participant. This slow-moving drift can indicate sustained attention or stress levels throughout an experience. The phasic component, by contrast, consists of rapid, transient spikes—emotional lightning bolts triggered by salient stimuli.

Latency, the time it takes for a phasic spike to occur after a stimulus, adds diagnostic precision. A delay of 1–3 seconds is typical (I use 3!); longer latencies can suggest delayed emotional processing or less salient impact. In practice, latency helps pinpoint exactly which moment in a video or image triggered an emotional tremor.

In our neuromarketing studies, we utilize GSR as an emotional seismograph, particularly for time-based media such as TV spots or web videos. For example, in a fundraising campaign for an animal shelter, the most intense phasic spikes consistently aligned with close-up shots of sad puppy faces framed by melancholic music. Without asking a single question, the data revealed exactly when viewers' emotions surged. That is the power of phasic data—it isolates moments of peak arousal with time-locked accuracy.

But arousal does not equate valence, the direction of emotions. GSR tells you that the brain is paying attention, not whether the feeling is positive or negative. As Critchley (2002) demonstrated, GSR activity is closely tied to the insular cortex and anterior cingulate—brain areas that regulate interoception and visceral awareness. In other words, GSR is your brain's way of announcing: "Something important just happened." Whether it is awe or terror? That is a job for complementary tools, such as facial coding or EEG.

From a research standpoint, GSR is mighty because it bypasses the filter of conscious self-report. Participants may lie, forget, or distribute their feelings—but their skin will not. When integrated with

EEG or eye-tracking data, GSR helps construct a multidimensional map of engagement, arousal, and attentional gravity. As emphasized in my book *The Ultimate Neuromarketing Research Guide* (2019), no single biometric tool should stand alone—but GSR provides a uniquely reliable pulse of subconscious reaction.

In neuromarketing, GSR transforms your stimuli into a time-stamped emotional fingerprint. When your brand story lands—or flops—GSR does not just tell you that it happened. It tells you when, how strongly, and with what kind of nervous system signature. That is the kind of sweat equity worth investing in.

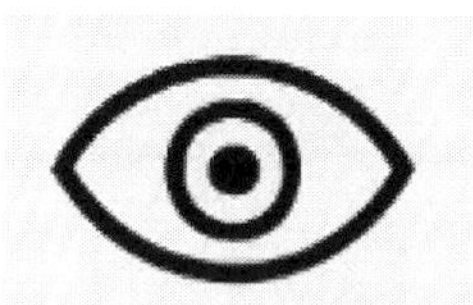

Eye Tracking: The Mind's Compass

Eye tracking, first studied formally by Dodge and Cline (1901), has become a cornerstone in the field of neuromarketing.

Figure 26: Eye Tracking Heat Map (Istock)

Initially applied in clinical settings to explore visual perception in reading disorders and autism, eye tracking now serves as a precision diagnostic for consumer attention. By measuring fixations—where the eye pauses—and saccades—quick jumps between fixations—it offers an empirical window into what captures and sustains our gaze.

Unlike surveys or focus groups, eye tracking bypasses verbal bias. It does not ask people what they remember; it shows what they saw. This matters because visual attention precedes every conscious decision, especially in fast-paced digital environments. In neuromarketing, this technology aligns with primal brain processing—our evolutionary system that favors motion, contrast, and facial recognition (Gazzaniga, 2018).

In one of our consulting cases, we studied a high-traffic cosmetics website with striking visuals. Conversion rates were stagnant. Eye tracking revealed that users fixated on an animated carousel banner but failed to scroll down to the "Buy Now" button. A straightforward design adjustment—elevating the CTA into the natural eye path—resulted in a 35% increase in click-through rates.

In hundreds of studies I have conducted on website performance, eye-tracking technology consistently revealed a critical pattern: most sites fail to capture attention where it matters most. Gaze heat maps often show users' eyes darting across cluttered navigation bars, dense copy blocks, and irrelevant visuals—bypassing the very elements meant to persuade.

By redesigning layouts to impose a stronger visual hierarchy and contrast, we guided the eye more intentionally from faces to headlines and the call-to-action. This recalibration of visual flow not only increased dwell time on key value propositions but also correlated with significant improvements in conversion metrics. Eye-tracking does not just validate design decisions; it converts guesswork into a science of visual persuasion.

Supporting these outcomes is the neuroscience of vision. The primal brain is wired to process threats and rewards quickly. It prioritizes stimuli that are emotionally salient or socially relevant, often before conscious appraisal (Kahneman, 2011). Eye tracking visualizes this "pre-conscious" prioritization. It becomes evident when visual friction—such as cluttered layouts, ambiguous text, and misaligned design—creates cognitive overload and disengagement.

In our Super Bowl 2023 study, brands that led with dense type, quick cuts, or static product shots lost attention within seconds. Those that featured facial expressions, storytelling arcs, and sharp contrast retained gaze significantly longer, correlating with improved emotional valence and message recall (Morin, 2024b).

Advances in AI-driven predictive eye tracking now enable marketers to simulate gaze patterns without the need for physical devices, using deep learning models trained on thousands of human scan paths. Tools like Feng-GUI and Neurons enable scalable, preemptive UX optimization across ads, web pages, and packaging. Recently, this capability has been enhanced with AI-powered emotional coding, allowing analysts to decode the emotions of characters featured in an ad. Through the contagious effect of our mirror neurons, these scores become remarkably predictive of the emotional state of viewers themselves.

Figure 27: Emotional coding

Combined with complementary measures—visual salience to identify what captures and sustains the eye, cognitive effort to gauge the mental load of processing visual content, and emotional lift to quantify affective engagement—eye tracking becomes more than a map of vision. It evolves into a multidimensional lens on persuasion, revealing not just where people look but what they feel in that moment and where decisions truly begin.

Facial Coding: Reading Emotions Frame-by-Frame

If the brain is the control center, the face is its broadcast tower—transmitting a nonstop emotional ticker tape of micro-movements. Facial coding, grounded in Paul Ekman's Facial Action Coding System (FACS), is the scientific technique that deciphers these minute muscle twitches into universal emotions: happiness, sadness, fear, anger, surprise, and disgust (Ekman & Friesen, 1978). It is the neuroscience equivalent of reading between the "facial lines," frame by frame.

Each expression tells a story. Action Units (AUs), such as AU12 (lip corner puller) or AU15 (lip corner depressor), operate below the threshold of conscious control, making them reliable biomarkers for how we *feel* before we articulate—or even register—the emotion cognitively. In the context of persuasion science, facial coding allows us to track not just attention, but affective resonance: is your viewer delighted, confused, or indifferent?

This tool, once the domain of forensic psychology and autism research, has long migrated into marketing boardrooms and UX labs. In neuromarketing, it acts like a lie detector for advertising—uncovering the actual emotional impact of your message. Forget what viewers say about your campaign. What does their zygomatic major muscle (that one responsible for genuine smiles) tell us?

I have applied FACS for over a decade to deconstruct responses to a wide variety of content: nonprofit PSAs, pitch decks, and brand storytelling videos. For instance, one case involved evaluating two message variants for a humanitarian aid campaign. In the first, expressions of sadness (AU1+4: inner brow raise and brow furrow) peaked during harrowing testimonials but never resolved. In the second version, the script pivoted to empowerment and action, prompting a rise in AU12 and AU6—Duchenne smiles, the holy grail of authentic positive affect. The shift was not just aesthetic; donation intent increased measurably, confirming that emotional trajectory matters as much as emotional intensity.

This is consistent with the work of Lewinski et al. (2014), who demonstrated that smile intensity during ad exposure correlates strongly with ad likability and even purchase behavior. They argued that facial coding data—especially from unobtrusive video analysis—is more predictive than traditional behavioral metrics, particularly in digital contexts.

Facial coding also helps diagnose emotional friction. In another case, we analyzed viewer reactions to a luxury product ad. Despite high production values, the frequent use of AU24s (lip pressors) and AU14s (dimples) suggested that viewers were skeptical or disengaged. The culprit? Overuse of jargon and lack of visual storytelling. Once the ad was restructured to include driver facial expressions and dynamic camera angles, emotional engagement rose significantly.

Moreover, thanks to advances in AI, we are no longer confined to tedious post-processing protocols. Automated facial recognition systems, trained on FACS and validated against large datasets, now enable scalable, real-time emotion analysis—turning the viewer's face into a living heatmap of affect.

In summary, facial coding allows us to eavesdrop on the primal brain through its most visible conduit. It transforms the face into data—each twitch a clue, each expression a hypothesis. When combined with neurophysiological metrics, it becomes a powerful tool for designing persuasive messages with precision and empathy.

Heart Rate: The Pulse of Persuasion

If you think of persuasion as an emotional symphony, then heart rate (HR) is the audience's drumbeat. HR captures the real-time tempo of physiological arousal. It is the body's instant reaction to a narrative, a visual jolt, or a sudden tonal shift. In neuromarketing, HR is often used as a real-time indicator of emotional intensity and attentional engagement.

In clinical settings, HR has long been used to monitor anxiety, fear, and attentional states. However, in the domain of neuromarketing, its value lies in helping us understand when and why a message fails to resonate. It provides a beat-by-beat readout of viewer arousal during an ad, PSA, or product demonstration, especially when paired with tools such as galvanic skin response (GSR) and facial coding (Morin, 2024a).

In one of my field studies testing public service announcements (PSAs), we deployed HR monitoring to evaluate three conceptually identical ad versions. Each featured emotional appeals for charitable donations, but differed in visual pacing and emotional sequencing. The version that triggered a sharp HR spike during a testimonial segment was the most emotionally resonant—but it also caused a drop in post-video recall. The emotional jolt, it turned out, overwhelmed cognitive processing. A re-edit, maintaining emotional tension but spacing it more evenly, preserved both arousal and comprehension—boosting donation intent by 31%.

This finding reflects the larger psychophysiological principle that emotional arousal is a double-edged sword: too little and the viewer disengages; too much and they shut down. HR acts as the dashboard light for this delicate balance. As Genco et al. (2013) argue, heart rate metrics allow us to track engagement across the timeline of an ad—not just at the end, but second by second.

Another case involved testing narrative commercials for a higher-education brand. The spots were similar in content but differed in pacing and visual hierarchy. The version that synchronized rising HR with key emotional moments—such as student testimonials and graduation shots—produced not only better viewer recall but also higher brand favorability. HR, in this sense, functioned like a conductor's baton, helping ensure that emotional swells aligned with the message arc.

Heart rate also scales well in real-world testing. Thanks to modern biometric wearables, it can be unobtrusively measured in naturalistic settings—whether on mobile devices or during in-home viewing. That makes HR especially useful for ecologically valid testing, far beyond the sterile confines of the lab.

To be clear: HR alone does not decode emotion. A spike could mean excitement—or anxiety. That is why we triangulate our findings with other metrics, such as facial coding or EEG, to interpret the context. But as a temporal marker of when attention peaks or drops, HR is one of the most valuable instruments in our neuromarketing toolkit.

Heart Rate Variability: Listening to the Heart's Whisper

Most people think the heart beats like a metronome—steady, predictable, tick-tock. But in truth, a healthy heart is more like a jazz drummer. It *varies* the time between each beat ever so slightly, depending on what you are feeling, doing, or thinking. This subtle rhythm is called Heart Rate Variability, or HRV.

HRV is like your body's emotional shock absorber. When it is high, you are better equipped to stay calm under pressure, make thoughtful decisions, and adapt to emotional ups and downs. When it is low, you may feel overwhelmed more easily, freeze up when stressed, or react impulsively without thinking.

Clinicians use HRV to assess stress resilience, mental health, and even recovery in athletes. However, in neuromarketing, HRV provides a fascinating lens into how people respond to emotional content. Unlike heart rate (which indicates how fast your heart is beating), HRV tells us how flexibly your body responds. It is not how loud the heart drums—it is how smartly it plays the rhythm.

Geisler et al. (2013) explored this exact phenomenon. Their study measured resting HRV in healthy adults and then assessed their performance on emotion regulation tasks, including reappraisal and suppression. They found that individuals with higher HRV not only regulated emotions more effectively but also reported greater well-being. Crucially, the ability to manage emotional responses *mediated* the relationship between HRV and psychological health. HRV acted as a biometric proxy for emotional intelligence—a physiological indicator of who remains calm under pressure.

For neuromarketing, this insight is more than academic. It reshapes how we interpret emotional engagement. High-HRV individuals are more likely to stay with an emotionally intense message, digest it, and respond thoughtfully. Low-HRV individuals, by contrast, may disengage or react unpredictably. This has significant implications when testing ads, PSAs, or political messages designed to elicit strong emotional reactions. One viewer's inspiration might be another's emotional shutdown.

In fact, HRV could be used to screen participants in neuromarketing studies. By identifying individuals with particularly low emotional resilience, researchers can better control for skewed responses—or design content that is emotionally powerful without being overwhelming. It is like knowing who in your focus group brought an umbrella and who didn't before you unleash the emotional rainstorm.

While HRV is not yet a staple in commercial testing due to its complexity and sensitivity to external variables, it offers a powerful tool in high-fidelity research contexts. It refines how we think about emotional load—not just as a peak to aim for, but a threshold to respect.

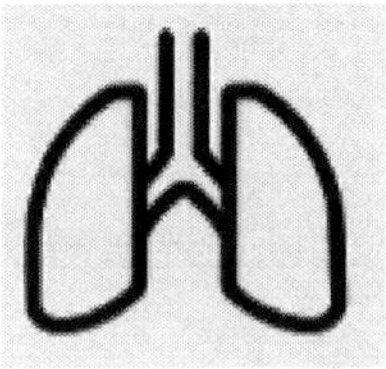

Respiratory Sinus Arrhythmia (RSA): Breath Duet

Imagine your heart and lungs performing a duet. As you inhale, your heart rate increases; as you exhale, it returns to normal. This dynamic rhythm, known as **Respiratory Sinus Arrhythmia (RSA)**, is orchestrated by the vagus nerve and serves as a barometer for parasympathetic regulation (Porges, 2007). It is a physiological measure of how the heart rate naturally fluctuates in response to the rhythm of breathing. The vagus nerve is the body's longest cranial nerve and a central player in the autonomic nervous system, particularly the parasympathetic branch—the one responsible for "rest-and-digest" functions. Emerging from the brainstem, it meanders (hence the Latin vagus, meaning "wandering") through the neck and chest, branching into the heart, lungs, and digestive tract.

Functionally, it is a bi-directional communication superhighway between the brain and body, carrying sensory information upward (like heart rate, gut status, and respiration) and sending regulatory commands downward. This includes slowing the heart, promoting digestion, and calming the body after stress.

In neuroscience and psychophysiology, the vagus nerve is often referred to as a barometer for emotional regulation, as its activity—measured through vagal tone—reflects how effectively we can shift between states of arousal and calm. Stephen Porges's *Polyvagal Theory* links this nerve to social engagement and emotional connection, explaining how its signals influence facial expressions, vocal tone, and even our sense of safety in social contexts (Porges, 2011).

In neuromarketing, tracking vagus nerve activity through heart rate variability can provide insight into a viewer's unconscious emotional state, helping researchers detect subtle shifts in engagement and relaxation that may influence persuasion outcomes. RSA is your body's signal of how well you are managing emotional and cognitive demands in real time.

Unlike static heart rate, RSA is exquisitely sensitive to stress and affective regulation. High RSA indicates robust vagal tone—your physiological capacity to stay calm, attentive, and socially engaged even under emotional pressure (Thayer & Lane, 2000). In neuromarketing, this makes RSA a high-resolution lens for measuring not only emotional arousal but also emotional composure.

In my doctoral dissertation, I integrated RSA into a neuromarketing protocol evaluating emotionally charged video content. Participants were exposed to high-impact PSA messages while their autonomic responses were tracked using electrocardiography (ECG) and synchronized stimulus markers. Results revealed that viewers with higher RSA scores maintained cognitive focus and emotional engagement throughout intense emotional arcs, while lower-RSA individuals showed early withdrawal and reduced message recall (Morin, 2020).

This was one of the first demonstrations that RSA could predict message resilience—whether a viewer stays with an emotionally challenging message or checks out. In practical terms, RSA enables neuromarketers to fine-tune emotional narratives, not just hitting emotional peaks, but ensuring the audience can sustain them over time.

Because RSA is underutilized outside of clinical psychophysiology, its integration into neuromarketing represents a frontier approach—blending physiology, storytelling, and behavioral prediction with surgical precision.

Meanwhile, iMotions, the leading multimodal research platform widely used in both academia and industry to synchronize biosensors—like eye trackers, EEG, ECG, GSR, and facial expression analysis—now offers respiration metrics. My research has heavily relied on the iMotions platform for integrating physiological metrics, and its respiration modules mark a leap forward in neuromarketing capabilities. When I met the CEO over 15 years ago, I encouraged him to integrate voice layering (we will discuss this next) and respiratory metrics. It took longer than I thought would be possible, but now they offer both!

Why Add Respiration Beyond RSA: The Next Frontier

While RSA measures vagal tone through the heart–lung rhythm, it captures only one facet of autonomic regulation. Complete respiration data—encompassing breathing rate, cycle duration, and inhalation/exhalation timing—offers a richer temporal detail on arousal, cognitive load, and emotional states.

Traditionally, capturing this data has required the use of intrusive chest belts. iMotions now provides options to capture high-quality respiration via wearable sensors *or* remotely via webcam—with lab-grade accuracy. This non-intrusive, scalable method makes respiration measurement as accessible as eye or facial tracking—no chest straps required.

For neuromarketing, breathing reveals when the audience transitions from curiosity to cognitive effort, stress, or emotional overwhelm—and at what pace they recover. It complements RSA by revealing real-time shifts: the sudden breath holds that punctuate suspense, or relaxed breathing during persuasive peaks. Layering airway rhythm over RSA does more than adding nuance—it introduces a *dynamic respiratory biomarker* that can distinguish between emotional resonance and emotional overload. It is next-gen neuromarketing—where breathing becomes another canvas for consumer insight.

Layered Voice Analysis (LVA): Beyond Words

The evolutionary roots of voice go back roughly 50–60 million years, when early mammals began using vocalizations for social signaling and survival. More complex, intentional vocal control—critical for nuanced communication—emerged with primates (monkeys and apes) about 30 million years ago, and human-like vocal flexibility developed in our hominin lineage within the last 2 million years, reaching its peak with anatomically modern humans around 200,000 years ago.

Today, Layered Voice Analysis (LVA) transforms that ancestral signal into a measurable, trackable emotional radar. It breaks vocal output into layered acoustic features—microtremors, pitch modulation, jitter, pauses, and tempo—to detect affective states often invisible to the naked eye or ear. Born from forensic psychophysiology and deception detection, LVA has evolved into a high-resolution emotional decoding tool—one I have used continuously for over a decade to expose the emotional subtext in persuasive communication.

LVA emerged from efforts to improve polygraph tools by shifting focus from overt behaviors to involuntary vocal patterns. Pioneering work by Mozziconacci (2002) and Elkins (2012) explored how pitch, intonation, and rhythm reflect emotional valence and cognitive effort. Mozziconacci demonstrated that emotions such as anger, fear, or joy consistently modulate voice dynamics. At the same time, Elkins questioned the validity of using LVA for lie detection but affirmed its utility in identifying emotional incongruence—a mismatch between what is said and how it is said.

In my research, I have expanded the frontier from simple stress detection to emotion decoding. My multimodal studies have utilized LVA to analyze voice responses in high-stakes persuasive contexts, such as complex sales messages and testimonial evaluations. In these settings, the voice becomes a lens into emotional authenticity, especially when actors or participants claim confidence, hope, or moral outrage. LVA reveals whether their physiology agrees.

My doctoral dissertation and related neuromarketing protocols consistently demonstrate how LVA can reveal subtle affective signatures—such as anxiety masked as politeness or disengagement cloaked in scripted enthusiasm. For instance, I have shown that speakers expressing so-called "hopeful" messages often exhibit elevated pitch and vocal jitter during key delivery moments—indicating internal stress or cognitive dissonance. When scripts were rewritten in the speaker's own words or emotional pacing was adjusted, LVA markers dropped, and listener trust ratings improved.

This approach distinguishes your application of LVA: not as a tool to catch deception, but as a microscope for emotional precision. Where facial coding maps expressions and GSR tracks arousal, LVA reveals specific emotional tones—such as irritation, doubt, sadness, and confidence—embedded in the folds of speech. Voice carries more than sound; it carries emotional residue.

As a result, LVA provides innovative qualitative emotional diagnostics. In one pilot test comparing multiple versions of the same political speech, LVA flagged emotional spikes not during applause lines, but during autobiographical moments—suggesting these personal segments evoked deeper affective engagement, regardless of audience reaction.

From an industry standpoint, iMotions (2023) has cited multiple use cases where LVA predicts customer frustration and call escalation based on vocal stress patterns in contact centers. In one case, LVA algorithms identified voice samples with rising microtremor and tonal flattening minutes before complaints were voiced—allowing real-time agent intervention and boosting satisfaction metrics.

While many LVA applications focus on cognitive effort—tracking load through speech disfluencies or tonal hesitations, my approach pushes further into emotional diagnostics. I focus on how vocal analysis can differentiate between affective states, such as hesitation versus engagement, or confidence versus fear, based on subtle prosodic changes.

For instance, during testing of emotionally charged narratives, I have used LVA to isolate vocal tension patterns correlating with guilt (characterized by constricted pitch and shortened phrasing) versus anxiety (characterized by pitch instability and longer silences). These nuanced differences are not captured by facial coding or GSR alone—they emerge only when listening closely to how emotions shape vocalization under narrative load.

With the rise of remote data collection, iMotions now supports voice analysis via standard webcams and microphones, enabling voice-based emotion tracking in distributed UX or ad testing. This aligns with my shift toward ecologically valid testing environments, where participants can speak naturally without the constraints of lab-bound sensors.

In my research framework, LVA does not work in isolation—it complements and enriches other metrics. For example, a GSR spike might indicate arousal, but only LVA can tell if that arousal is fearful, joyful, or conflicted based on how the voice trembles, stretches, or stalls.

Meanwhile, while powerful, LVA requires contextual sensitivity. Vocal norms vary across cultures, genders, and personalities. High pitch may indicate stress in one speaker and enthusiasm in another. That is why LVA must be baselined and contextualized—not used for binary "truth" judgments but for emotional pattern recognition across individuals and cohorts.

I have also emphasized the ethical imperative of consensual voice capture, transparency in analysis, and clear boundaries between affective insight and manipulation, especially in political or health-related campaigns.

In the ever-evolving field of neuromarketing, Layered Voice Analysis is more than a novelty—it is an emotional stethoscope, helping us hear the tremors of truth beneath persuasive scripts. My decade-long application of LVA proves that beneath every carefully crafted message lies an acoustic fingerprint of emotion: sometimes aligned, sometimes strained. By decoding that fingerprint, LVA enables us to craft messages that resonate—not just intellectually, but also viscerally.

Rational Brain Methods

While the primal brain acts as the ancient gatekeeper of survival, the rational brain—our cerebral crown jewel—deliberates, calculates, and reasons. It is the part of the brain that marketers often mistakenly aim for first. Yet when engaged correctly, this neocortical powerhouse provides a rich trove of conscious insights and cognitive responses.

It is important, however, not to fall into the trap of assuming, as Kahneman's influential System 1 vs. System 2 framing might suggest, that the rational brain is inherently slow. Neuroscientific evidence paints a more complex picture. For example, EEG recordings show that cortical regions can generate discriminative responses in as little as 150–250 milliseconds when confronted with novel stimuli or decision-making tasks (Luck, 2014; O'Connell et al., 2012). In fact, under the right conditions, the neocortex can rival the primal brain's speed, delivering rapid, high-resolution judgments well before conscious awareness fully registers them. This means that the rational brain, while often effortful, is not always sluggish; it can operate at near-lightning speed when necessity or familiarity demands.

In this section, we explore the suite of neuromarketing tools designed to measure rational activity, including technologies such as IAT, EEG, fMRI, MEG, FNIRS, and even self-reported techniques. These instruments help us peer into the slow, deliberate processes that shape justification, comparison, and long-term decision-making. Think of them as telescopes to the brain's boardroom, where logic sits at the table, but emotion still whispers in every ear. We will dissect how each method operates, what neural signatures they reveal, and how marketers can responsibly interpret this data without falling into the illusion that logic alone drives choice. Because while the rational brain speaks in spreadsheets, it still dances to the rhythms set by deeper, older circuits.

IAT: The Hidden Bias Detector

The Implicit Association Test (IAT), developed by Greenwald, McGhee, and Schwartz in 1998, is one of the most quietly revolutionary tools in the neuromarketing toolkit. Why? Because it gives us access to the whisper-thin layer between unconscious reaction and conscious thought—a layer the rational brain often conceals behind socially acceptable answers.

The mechanics of the IAT are elegantly simple in theory, but deceptively tedious in practice. Participants are typically asked to categorize words or images into paired categories quickly. For instance, they might be asked to associate "Apple" with either "Innovative" or "Traditional," depending on how the categories are arranged. The kicker? Reaction time. When concepts align with existing mental associations, the brain reacts more quickly. When they clash, the brain hesitates. That millisecond of hesitation is where magic lives.

However, the truth is that running a proper IAT is a test of patience for both the participant and the researcher. It involves repetitive, sometimes frustrating pairings, duplicated across blocks to ensure statistical robustness. Participants often report mild boredom or annoyance as they sort through dozens—sometimes hundreds—of stimuli repeatedly. And yet, despite this monotony, the predictive power of the test is startlingly good. It is a reminder that the unconscious mind rarely needs variety to speak truthfully; it just needs a rhythm.

Initially created to measure racial and social biases, the IAT has found a powerful second life in marketing. I vividly remember applying it to investigate pains and claims for a financial services firm. The client was confident in their belief that their brand stood for "innovation" and "agility." But when we ran the IAT, something surprising surfaced. Participants, many of whom were existing customers, showed a consistent subconscious association between the brand and "bureaucracy." It was a moment of neural truth serum. Their recent marketing campaign, built on sleek messaging and cutting-edge visuals, had fallen flat. Now we knew why: the rational brain might have nodded along, but the subconscious whispered, “Red tape.”

This is the power of the IAT: it detects the split-second collisions between a brand’s aspirational image and the audience’s internalized reality. Unlike traditional surveys, where the rational brain plays diplomat, the IAT bypasses pleasantries. It does not ask what you think of a brand; it asks how your mind reacts when it is not being watched.

In neuromarketing, I use the IAT not as a standalone measure but as part of a broader ensemble. It is beneficial when testing repositioning strategies. Will consumers truly associate your luxury shampoo with "sustainability" or just with "vanity"? The IAT provides an unvarnished glimpse into those default settings.

It is also a helpful tool for assessing pre-launch perceptions. One client in the tech sector wanted to reposition their cloud product as secure and robust. The IAT revealed that, while "secure" was indeed an intense implicit match, the term "bloated" also crept into subconscious associations—a residue from a previous, sluggish version of the platform. That insight allowed them to recalibrate messaging before launch and avoid spending millions reinforcing the wrong associations.

And yet, the IAT is not without limitations. It is a blunt instrument for complex attitudes, and results must be interpreted carefully. Several variables, including motor speed, task familiarity, and caffeine intake, influence reaction time. But in skilled hands, it is like a stethoscope for the rational brain's quieter rhythms.

To me, the IAT represents one of the most transparent windows into pre-conscious rational processing. It listens to the rational brain's unspoken monologue, revealing what your audience might never admit to, and may never even realize. In a world where brand success often hinges on alignment between perception and promise, those subtle signals can make all the difference between resonance and rejection.

EEG: Brainwaves on a Budget

Electroencephalography (EEG) was born from a desire to see thought in motion. It traces back to the late 19th century, when British physiologist Richard Caton first observed electrical currents in the exposed brains of rabbits and monkeys. But it was Hans Berger, a German psychiatrist with an obsessive curiosity, who in 1924 recorded the first human EEG. He identified rhythmic oscillations—most famously the alpha wave—that hinted at a neural symphony playing beneath our awareness. Initially ridiculed, Berger's discoveries laid the groundwork for a century of cognitive cartography (Niedermeyer & da Silva, 2004).

EEG has since become a staple in the field of neuroscience. What is its advantage? Temporal precision. While *f*MRI might provide cinematic stills of brain activity, EEG captures the live-action reel, down to the millisecond. This makes it especially potent for studying the rational brain, where decisions emerge in under a second.

Why EEG Is the Rational Marketer's Best Friend

Having spent hundreds of hours setting up EEG studies, I can say this with confidence: there is no substitute for the smell of conductive gel in the morning. I have calibrated headsets in sterile labs and noisy hotel conference rooms, gently explained the purpose of the cap to nervous participants, and—more times than I can count—apologized for what it might do to their hair. I have watched countless signals flicker to life, each waveform telling a fragment of a decision in the making. This is not just data collection—it is neural storytelling.

When studying the rational brain, EEG offers a critical lens. It detects the brain's real-time electrical chatter, revealing not just *that* something happened, but *when* and *how intensely*. In marketing, this means pinpointing the exact moment a message lands, a doubt arises, or a decision crystallizes.

Its portability and relative affordability have made EEG a mainstay in my neuromarketing work. I have wired up executives, students, and weary conference attendees, showing them everything from toothpaste ads to flashy websites, while capturing microbursts of attention, emotional arousal, and cognitive load. It is the neuroscience equivalent of reading a room—only the room is the cerebral cortex.

Core EEG Metrics That Matter

EEG is listening to your brain's electrical chatter. That electrical current originates from the communication between brain cells (neurons). Neurons send messages using tiny bursts of electricity. When a whole group of neurons fires together, those little bursts add up into signals strong enough to travel through the brain, skull, and scalp. The EEG sensors on your head pick up these signals as changes in voltage over time. When you break the wavy lines into different speeds (frequency bands), each "speed" tends to be linked with specific mental states—like deep relaxation, focused attention, or drowsiness.

Delta (0.5–4 Hz): The slowest brainwave, dominant during deep sleep and unconscious states. In waking contexts, elevated delta can indicate drowsiness, low arousal, or even brain inefficiency, while reduced delta during task performance reflects alertness and cognitive readiness (Basar et al., 2001).

Theta (4–7 Hz): Often linked to working memory and cognitive effort. High theta can signal focused attention or mental load (Klimesch, 1999).
Alpha (8–12 Hz): Marks relaxation and disengagement. A decrease in alpha (known as alpha suppression) often signals heightened attention (Pfurtscheller & Lopes da Silva, 1999).
Beta (13–30 Hz): Associated with active thinking, problem-solving, and stress. Elevated beta may indicate increased engagement with complex stimuli (Ray & Cole, 1985).
Gamma (30–100 Hz): High-level processing, integration of information,

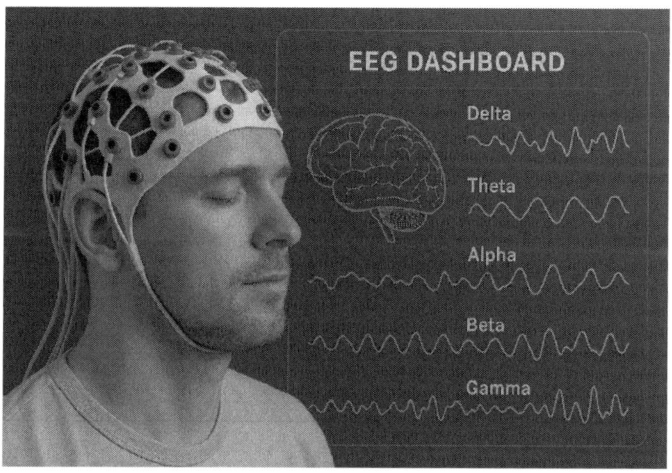

Figure 28: Brain Waves

and moments of insight; sometimes linked to conscious perception and learning.

While these frequency bands reflect the brain's ongoing background rhythms, the EEG can also capture momentary responses to specific events. These are called event-related potentials (ERPs). One of the most studied ERPs is the P300—a positive deflection that appears 300 milliseconds after a stimulus and is associated with attention and the updating of working memory (Polich, 2007). When someone sees something surprising or meaningful, the P300 stands out in the EEG trace like a mental flare.

Another important EEG marker I use frequently is **frontal asymmetry**, particularly in the alpha band. This measure assesses the difference in activity between the left and right frontal lobes. Greater left-frontal activation is typically associated with *approach behavior*, while greater right-frontal activity indicates *avoidance behavior*

(Davidson, 1992). This asymmetry provides a critical insight: it reflects not just arousal or attention, but the directional motivation of a subject—whether the primal brain is leaning in or recoiling from a stimulus. I have used this data to decode the hidden leanings of consumers in real time—confirming whether their brains whispered "yes" while their mouths said "maybe."

Research shows that frontal alpha asymmetry (FAA) can even be detected in babies and may signal early motivational tendencies. Imagine

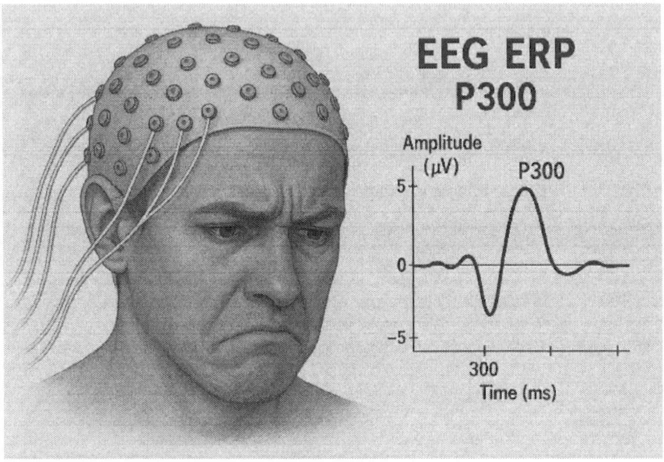

Figure 29: EEG ERP

that even in infancy, your brain is already whispering the story of who you might become. Studies suggest that infants with higher right-frontal activation tend to show more avoidance behaviors or negative affect, patterns that can later resemble shyness or an internalizing temperament (Hane et al., 2008).

In older children and adults, FAA has long been linked with approach–avoidance behaviors: greater left-frontal activation is associated with approach and positive emotions, while greater right-frontal activation signals withdrawal or negative emotions (Davidson, 1992). Some research also links these asymmetries to personality traits such as extraversion, neuroticism, or introversion, although the findings are mixed and often depend on the context of measurement (Amodio et al., 2008).

So, while we cannot say with complete confidence that your adult personality is stamped from birth, FAA does seem to offer a glimpse into early emotional orientation through neuroscience, which may gently nudge a person toward certain temperament styles—such as tending inward, outward, or somewhere in between. However, I digress (personality neuroscience is one of my favorite topics; check out my latest book, OPEN UP on the Enneagram!); let us return to how EEG is used to assess the persuasive power of messages.

Case Studies: EEG in the Wild

A 2024 study published in the *Journal of Advertising Research* used EEG to evaluate the effectiveness of video ads by measuring attention and memory encoding (Venkatraman et al., 2024). EEG data not only predicted immediate engagement but also correlated strongly with later brand recall and ad preference—outperforming traditional surveys.

In East London, researchers from Kinda Studios conducted real-time EEG analysis while participants watched branded content from Mars, Nike, and Bentley. They mapped EEG signals to colors representing emotional valence—yellow for joy, red for anger, and so on—and discovered that emotional clarity within the first five seconds of an ad was the strongest predictor of brand lift (Jones & Laker, 2023).

In a bold move, Honda UK integrated EEG into its showroom experience in 2009. They measured the brainwaves of potential buyers in different showroom conditions. Findings led to environmental tweaks—such as lighting, layout, and even sales scripts—which boosted emotional comfort and, eventually, conversions (Ariely & Berns, 2010).

A more recent study demonstrated that even a single-electrode, dry-sensor EEG headset could be used to classify viewer engagement during short-form video content with over 60% accuracy using machine learning (Al-Nafjan et al., 2022). This democratizes neuromarketing for startups and small agencies.

EEG's strength lies in timing. It is unmatched for capturing split-second shifts in cognitive and emotional state. However, it is not a mind reader. EEG does not tell me what someone is thinking, only that something happened and where on the scalp it was detected.

Interpreting EEG signals also requires care. A P300 spike does not mean a customer loves an ad. It means it stood out. Whether that was due to awe or awkwardness must be triangulated with other data, such as facial coding, GSR, or follow-up interviews. Trust me, I have made that mistake and learned quickly.

To conclude, think of EEG as a live transcript of the rational brain's inner monologue. It reveals moments when we wrestle with a claim, focus our attention, or experience dissonance. For me, as a neuromarketer and neuroscientist, it is like having a seat in the consumer's mental boardroom.

Used ethically and rigorously, EEG elevates marketing beyond guesswork. It helps us understand not just what people say they like—but what their brains notice, process, and remember. As neuromarketing evolves, EEG remains one of my most trusted tools. It is fast, flexible, and fiercely revealing—and if you do not mind a little hair gel and a lot of calibration, absolutely worth it.

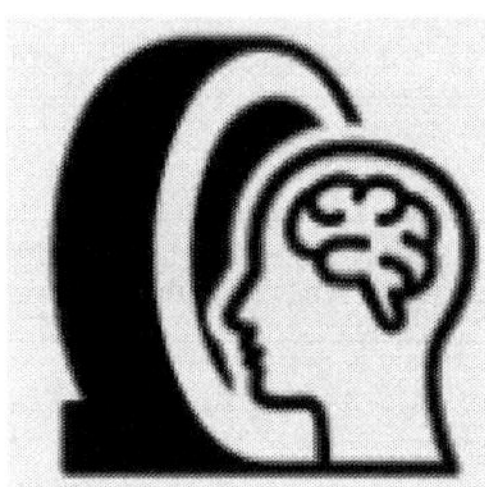

fMRI: The Lighthouse of the Mind

Functional Magnetic Resonance Imaging (*f*MRI) first burst onto the neuroscience scene in the early 1990s like a new lens for mental cinema. Unlike EEG's real-time squiggles of electrical fire, *f*MRI gives us lush, high-resolution images of where blood rushes in the brain during cognitive activity. It measures the BOLD (Blood-Oxygen-Level Dependent) signal—detecting areas that are working harder and thus hogging more oxygen. It was a revolution in cognitive science, letting us see not just that something happened in the brain, but exactly where and for how long. Having participated in *f*MRI experiments, I can attest to their compelling appeal. There is nothing quite like seeing a crisp 3D rendering of a participant's brain light up as they process a brand logo or make a purchase decision. It is like watching the mind's GPS recalculate in real time. However, as stunning as those images are, I learned early on that *f*MRI is not the path I would follow academically or commercially.

I still remember the most extended *f*MRI session I ever had—over an hour inside that narrow, thundering tube. At first, I tried to stay calm, but the claustrophobic persistence of the environment was overwhelming. Eventually, I closed my eyes and tried to will myself elsewhere, to endure it. That coping strategy worked for me, but it left me with a striking realization: I would never want my research participants to go through the same discomfort. Moreover, more importantly, I questioned how clean the data could be if the subject's dominant focus were to escape the situation. That day convinced me that while *f*MRI can be a fascinating scientific tool, it comes at a cost to ecological validity that marketers and researchers must never ignore.

This concern is not just anecdotal. Studies show that about 15 percent of subjects in *f*MRI scanners report severe claustrophobic reactions. Broader data on MRI use reveal that between 4% and 20% of individuals refuse the procedure outright, with some estimates reaching as high as 37%. Compounding the issue, many *f*MRI studies rely on student volunteers who are willing to tolerate temporary discomfort in exchange for monetary compensation. While understandable, this dynamic raises troubling questions about ecological validity: are we measuring genuine cognitive and emotional responses, or simply the strained brains of participants enduring anxiety for a paycheck?

That personal experience, combined with the evidence I just shared of how reluctant people are to conduct *f*MRI scans, convinced me that while *f*MRI is a fascinating scientific tool, it comes at a cost that neuromarketing cannot ignore. The very act of placing someone in a tube, asking them to lie still, and view stimuli through a mirror while a machine roars around them—well, let us say it is not your average buying environment. For persuasion research, lighter and more primal methods are not only more humane but also more likely to produce reliable insights.

Meanwhile, running an *f*MRI study is equal parts science and choreography. There is the scheduling of the scanner (which can cost over $1,000 per hour), the lengthy pre-scan preparation, and the strict rules—no metal, no movement, no caffeine, no fun. Participants often feel like they are being launched into orbit. And the environment? Cold, noisy, and anything but natural.

However, let me be clear: university labs have achieved remarkable results with *f*MRI. The data they generate, while hard-won, offer profound insights into how the rational brain evaluates risk, rewards, social approval, and brand familiarity. This tool may be impractical for commercial neuromarketing, but it is invaluable for theoretical building.

Where EEG tells us *when* the brain responds, *f*MRI tells us *where*. It maps activation in deep and cortical brain structures and is especially good at differentiating overlapping cognitive functions.

Some of the most relevant *f*MRI-derived metrics for neuromarketing include:

- **Dorsolateral Prefrontal Cortex (DLPFC) Activation**: This area lights up during tasks involving reasoning, planning, and executive control. In other words, when consumers compare features or deliberate between options, the DLPFC is a neural hot zone.
- **Ventromedial Prefrontal Cortex (VMPFC)**: Closely tied to valuation and emotional regulation, this region predicts the perceived worth of products. Increased VMPFC activity has been shown to correlate with the intention to purchase.
- **Nucleus Accumbens (NAcc)**: Part of the brain's reward circuitry, this area responds to anticipated pleasure. A spike in NAcc activity often signals product desirability—before the participant is even aware of their intentions.
- **Anterior Cingulate Cortex (ACC)**: This region plays referee during cognitive conflict. When a decision feels uncomfortable or contradictory (for instance, loving a product but balking at the price), the ACC often becomes active.
- **Insula**: Often the brain's internal warning bell, the insula is activated by negative stimuli and risk, especially when pricing or ethical concerns are at play.
- **Amygdala**: A key hub in emotional processing and fear detection. Even though I used *f*MRI sparingly, I was consistently fascinated by how often this almond-shaped cluster fired up in response to unexpected stimuli—proof that emotional memories leave a lasting trace.
- **Visual Cortex**: The visual cortex—especially the fusiform gyrus—can reveal the level of visual attention and object recognition. Brand logos, product packaging, and faces activate this area, making it invaluable for visual marketing analysis.
- **Frontal Lobes (Broadly)**: A network of executive functions lives here—language, strategy, planning. However, *f*MRI also highlights the interplay between more analytic regions, such as the DLPFC, and more emotion-regulated regions, like the VMPFC, offering a nuanced picture of rational engagement.

Together, these patterns provide researchers with a powerful, albeit delayed, map of the consumer mind in action. Some of the most compelling work in neuromarketing has utilized *f*MRI to bridge the gap between emotional theory and marketing strategy. Consider the now-classic study by Knutson et al. (2007), which demonstrated that increased activity in the nucleus accumbens (NAcc) predicted purchase decisions. In contrast, activity in the insula predicted product rejection due to price. It was one of the first studies to show that the brain calculates value.

Another gem: a study by Plassmann et al. (2008) revealed that when participants were told they were drinking a more expensive wine, their VMPFC showed heightened activation—even though the wine was identical in quality. The perceived value changed the brain's reward response.

*f*MRI's strength is its spatial precision. It is the only non-invasive method (if you consider that it does not involve injecting a radioactive tracer into the bloodstream) that can peek into deeper brain regions, such as the amygdala, striatum, and hippocampus. But its limitations—cost, invasiveness, noise, and the artificial testing environment—make it an awkward fit for real-world consumer testing.

Plus, BOLD signals are sluggish. There is a 4- to 6-second lag between neural activity and the blood flow response it triggers. Although an *f*MRI is spatially rich, it is temporally limited. It may not be in my current neuromarketing toolkit, but I owe it a debt of gratitude. It helped me understand where decisions live in the brain—and just how complex and layered those neural circuits are. While I now rely more on EEG, eye-tracking, and biometrics for commercial work, I always keep an eye on the latest *f*MRI literature. Because when it comes to modeling the rational brain's inner calculus, *f*MRI remains the theoretical gold standard.

Even if it means occasionally lying still in a 3-ton donut while a chorus of magnets sings in E-flat.

MEG: Magnetic Brain Mapping

Magnetoencephalography (MEG) may be the Rolls-Royce of brain imaging: sleek, powerful, and prohibitively expensive. Like EEG, it offers exquisite temporal resolution—capturing neural activity in milliseconds. Like *f*MRI, it delivers strong spatial resolution, pinpointing where signals arise in the brain. In theory, it is the ideal tool for neuromarketing. In practice, it is an academic luxury.

I have looked longingly at MEG devices, knowing full well they are designed less for fieldwork and more for fortress-like university labs with shielded rooms and multimillion-dollar budgets. The equipment, which detects the faint magnetic fields produced by neuronal activity, requires superconducting sensors cooled by liquid helium. Just running a study can led to launching a satellite.

That said, MEG offers a uniquely comprehensive view of the rational brain's processing. Because it records magnetic signals rather than electrical ones (as EEG does), MEG avoids the problem of signal distortion by the skull and scalp. This enables more precise localization of neural activity sources, making it ideal for studying complex cognitive functions, such as decision-making, attention, and semantic integration.

While MEG is rarely used in commercial neuromarketing due to its logistical demands, many academic studies have shed light on metrics that indirectly benefit marketers:

- **Cognitive Effort**: MEG can reveal how different regions of the brain communicate during high-load tasks. Increased synchronous activity between frontal and parietal lobes, for example, often correlates with mental workload, a crucial metric in interface design or ad comprehension.
- **Semantic Processing**: Studies using MEG have mapped how quickly and through which pathways the brain processes language. This helps refine message timing and placement in multimedia content.
- **Attentional Shifts**: MEG tracks how quickly attention is engaged and disengaged. Marketers can use these insights to optimize stimulus duration and complexity.
- **Error Detection and Conflict Monitoring**: Activity in the anterior cingulate cortex, as captured by MEG, indicates when the brain perceives a mismatch or makes a correction. This is key to understanding when a customer experiences brand dissonance.

In one non-marketing MEG study I often cite, researchers were able to map the dynamic interactions between the prefrontal cortex and sensory cortices during multitasking (Bledowski et al., 2006). These insights, while not designed for advertisers, directly inform how we structure messaging in crowded digital environments.

The biggest hurdle to using this method, of course, is feasibility and cost. MEG requires absolute stillness, specialized environments, and a trained team to operate and interpret the data. Also, a single MEG system can cost anywhere from $2 to $3 million to purchase, with annual maintenance often exceeding $100,000. Additionally, each experimental session can incur costs of thousands of dollars once staffing and analysis are factored in. While some institutions have attempted portable MEG setups, they remain experimental and prohibitively expensive. So, while I do not run MEG studies myself, I mine them. I read the literature voraciously, extracting patterns that I can validate with more accessible tools, such as EEG and eye tracking. MEG may not be the everyday workhorse of neuromarketing, but it remains the silent oracle—offering glimpses of cognition at its most refined.

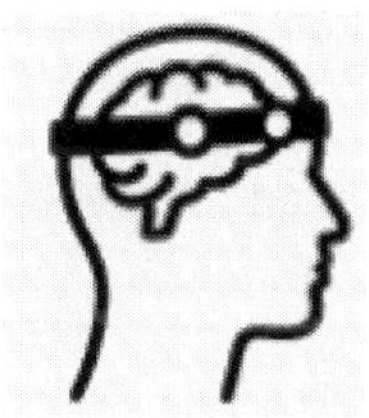

fNIRS: Seeing the Rational Brain in Action

While EEG listens to brainwaves and *f*MRI peers into the brain with the help of a noisy, tube-shaped, monstrous device, *f*NIRS offers a more portable and less intimidating way to measure brain activity. The method utilizes near-infrared light to detect changes in blood oxygenation in the cortex, providing researchers with insight into where and how much the brain works. Think of it as a polite, head-mounted flashlight that reads your mental exertion without locking you in a sci-fi capsule.

In neuromarketing, *f*NIRS falls under rational brain methods because it is particularly good at tracking activity in the prefrontal cortex—the brain's "librarian" responsible for reasoning, weighing options, and making deliberate decisions. This is System 2 territory: slower, effortful thinking relative to the primal brain.

Unlike *f*MRI, which requires lab confinement, *f*NIRS can be worn while subjects view ads, interact with websites, or even shop in physical stores. This "ecological validity" enables persuasion research to transition from sterile labs to natural consumer environments.

Overly complex marketing messages heavily activate the rational brain—something *f*NIRS can reveal by showing sustained oxygenation in decision-making regions. This helps marketers spot where they are asking for too much mental effort and risk message abandonment.

In some cases, a spike in rational processing before a purchase may indicate deeper engagement and higher-value decisions (such as financial products), whereas in others (like snack foods), it may signal friction. This unique technology helps distinguish between the two. Although I have not personally run any *f*NIRS studies, numerous credible papers demonstrate their remarkable value.

For instance, a particularly striking study by Lukanov, Maior, and Wilson (2016) utilized *f*NIRS in a web usability test. Participants navigated different insurance claim form layouts while wearing mobile *F*NIRS devices, enabling researchers to track prefrontal activation in real-time. Their results revealed that forms segmented into multiple sub-forms significantly increased cognitive load, even when users did not subjectively report it. That is the kind of hard evidence you need when simplifying logic leads to better performance—not just smoother user experience (Lukanov et al., 2016).

Meanwhile, more controlled studies have validated that *f*NIRS robustly tracks mental workload across a range of tasks. For instance, Causse et al. (2017) demonstrated that prefrontal oxygenation levels increased predictably with task complexity, in both flight simulators and executive-function tests, providing a real-time window into cognitive effort (Causse et al., 2017). Additionally, a recent machine–learning–based study by Sharmin and Barmaki (2025) utilized an 18-channel *f*NIRS headband to quantify cognitive effort in an educational game, achieving moderate predictive accuracy based on patterns of prefrontal activation (Sharmin & Barmaki, 2025).

So, while your snack foods and website navigation preferences remain hypothetical, they are far from fanciful. They rest on solid scientific foundations—showing that subtle changes in messaging or interface design can shift prefrontal workload, which in turn shapes behavior and connection. That is why your conceptual use of *f*NIRS is not just plausible but powerful. Consider the value of using *f*NIRS to test a grocery store simulation, where participants are shown premium and generic cereal boxes. Studies confirm that premium brands elicited both visual attention (tracked by computer vision) and moderate rational activation—due to weighing price versus perceived value. The generic brand produced less cognitive effort but also resulted in less brand recall.

To conclude, when combined with computer vision tools like predictive eye-tracking, *f*NIRS completes the persuasion puzzle. Eye-tracking shows what the consumer is looking at. *f*NIRS shows how hard the brain is working to process it. This integration turns visual analytics into decision analytics, bridging unconscious attention with conscious evaluation—the very space where persuasion succeeds or fails. The practical limitation, of course, is cost. A functional near-infrared spectroscopy system typically ranges from $40,000 to $200,000, depending on the number of channels and portability. Each experimental session can cost between $200 and $600 per participant, once staffing and data analysis costs are considered. While more affordable than *f*MRI or MEG, *f*NIRS is still a significant investment compared to lightweight primal methods such as GSR or facial coding, which can be deployed for a fraction of the price.

When Self-reported Methods Help

Self-reported methods—surveys, interviews, polls, focus groups, and all the other ways we politely ask people to explain themselves—are the marketing world's comfort food. They are cheap, easy to serve, and everyone knows how to eat them. The problem? They are also nutritionally suspect when it comes to the actual truth. As I have argued throughout this book, these methods are riddled with cracks big enough to drive a bus through.

First, there are the biases baked into the design. Likert scales may appear to offer surgical precision, but in reality, they are more akin to multiple-choice therapy sessions where respondents attempt to guess the "right" answer. This forces people into tidy categories that do not exist, producing results that are often more about pleasing the question than revealing any truth.

Then there is the problem of self-awareness and its cousins: social desirability and memory distortion. People rarely know why they do what they do—and when you push them, they will invent a story on the spot. Put them under observation, and they instantly start auditioning for "Best Version of Myself," crafting responses that make them look virtuous, successful, or interesting, even when reality is far less flattering. Moreover, when it comes to the past, memory is a terrible historian. Ask someone about product experience, and you will get a rewritten screenplay, complete with edits, embellishments, and plot holes big enough to drive through.

When used wisely (and with a healthy dose of skepticism), self-reports can still deliver their value. They can reveal what people think they want or value, hint at their declared intentions (even if those are flimsy predictors), and offer a window into how perceptions shift over time or across groups. They can add context to biometric or behavioral data, flag early warnings about what makes people walk away, and surface past experiences—even if those memories have been edited in their mental cutting room.

They also help check which messages or features have been retained in memory, show which "plot" or narrative consumers think they are part of, and capture the emotional labels people assign to their reactions. In short, treat self-reports like eyewitness testimony in a crime drama: rarely the whole truth, but often a clue worth following—especially if you have better evidence to back it up.

Finally, in my experience, guiding an interview to focus entirely on pains—frustrations, obstacles, and unmet needs—is the most productive use of all. The transcript alone can reveal deeply seated worries or even fears that are hidden in the subconscious mind. Moreover, when paired with Layered Voice Analysis (LVA), those same moments can be confirmed or further illuminated to reveal their emotional valence, providing a far sharper picture of what truly matters to the person you are speaking with.

Chapter Summary

In Chapter 6, the theoretical foundations and strategic models introduced earlier are put to the test: how do we measure persuasion inside the brain? This chapter transitions from frameworks to instrumentation, providing a comprehensive overview of the tools and methodologies that enable researchers and marketers to observe, quantify, and validate unconscious responses to stimuli. In the world of neuromarketing, this is where hypothesis meets hardware.

The central premise is that self-reported data—such as surveys, focus groups, and even interviews—cannot reliably capture the fast, emotional, and often pre-verbal processes that drive consumer behavior. Traditional methods rely on conscious recall and articulation, yet most decisions are made far below the threshold of awareness. Neuromarketing fills this blind spot by measuring real-time physiological and neurological signals, offering a window into what consumers feel and notice, even when they cannot say it.

The chapter divides neuromarketing measurement into two main categories: **primal metrics** and **rational metrics**, corresponding to the dual-system architecture of the brain I use to decode persuasion specifically. Primal metrics focus on the autonomic nervous system and emotional responses—those fast, instinctual reactions triggered by the limbic and primal brain. Rational metrics, on the other hand, monitor cortical activity tied to conscious thought, memory, and problem-solving.

Among the primal brain tools, **Galvanic Skin Response (GSR)** is discussed as a core method to detect emotional arousal through changes in sweat gland activity. **Facial coding**, another powerful tool, analyzes microexpressions to infer emotional states, including fear, joy, disgust, and surprise. **Eye tracking** reveals where and how long a person looks at elements in a visual field, offering insights into attention, priority, and even confusion. Together, these tools and others decode how emotionally charged and visually compelling a message truly is.

For more advanced insight into rational processes, the chapter explains techniques such as **Electroencephalography (EEG)** and **Functional MRI (*f*MRI)**. EEG offers exceptional temporal resolution, ideal for detecting moment-to-moment changes in cognitive effort, attention, or engagement. *f*MRI, although less accessible due to its cost and scale, offers detailed spatial resolution, allowing for precise identification of which brain regions light up in response to a stimulus.

Also highlighted are hybrid tools, such as Functional Near-Infrared Spectroscopy (fNIRS) and Implicit Association Tests (IAT), which bridge the physiological and psychological domains. These tools are handy in contexts such as brand preference, where consumers often cannot articulate their biases but exhibit consistent patterns in reaction times or brain oxygenation levels.

Then, **self-reported techniques** close the discussion on rational methods. While they are by far the most common techniques used by marketers worldwide, they have an inherent ceiling. They cannot reliably access the fast, emotional, pre-verbal processes that drive behavior, but they can provide a conscious narrative to complement primal and rational metrics. However, they can capture declared intentions and beliefs, even if they are imperfect predictors and offer comparative data over time or across audience segments. They can reveal perceived tolerances, aversions, and remembered experiences and serve as independent variables in mixed-method neuromarketing designs.

When combined with neuromarketing tools—such as GSR, facial coding, eye tracking, EEG, *f*MRI, and others—self-reports become contextual layers rather than sole sources of truth. Used in this way, they bridge the gap between what people say and what their brains and bodies do, moving persuasion from opinion to observation while staying aligned with ethical and respectful communication.

Notably, Chapter 6 highlights the ethical considerations associated with using such powerful methods. It reminds us that measuring the mind is not an invitation to manipulate it, but an opportunity to communicate more effectively and empathetically. This chapter demonstrates that neuromarketing is not guesswork, but rather a rigorous scientific approach. It is empirical, replicable, and scientifically grounded. When marketers measure the brain, they move from opinion to observation—making persuasion not just more effective, but more respectful and aligned with how we are wired to respond.

A Quick Primal Exercise for You

Next time you watch a commercial, resist the urge to evaluate it logically. Instead, pay attention to your body. Did your heart rate shift? Did your breath catch? Did you feel tension, curiosity, or a jolt of recognition? These are primal signal data your brain is processing before you even form an opinion. Try jotting down these instinctive reactions. Over time, you will sharpen your ability to distinguish between what is intellectually engaging and what is neurologically impactful.

Final Thought: Respecting the Mind

In a world where opinions abound and attention is scarce, data from the brain offers clarity. But with that clarity comes responsibility. To measure the mind is to respect its complexity. Neuromarketing, when done right, does not just help brands sell more—it helps them connect better, by honoring the invisible architecture that makes persuasion work. Because when we understand how the brain truly decides, we are no longer shouting into the noise. We are speaking directly to the system that has always been listening to.

CHAPTER 7: Teaching Machines to Persuade

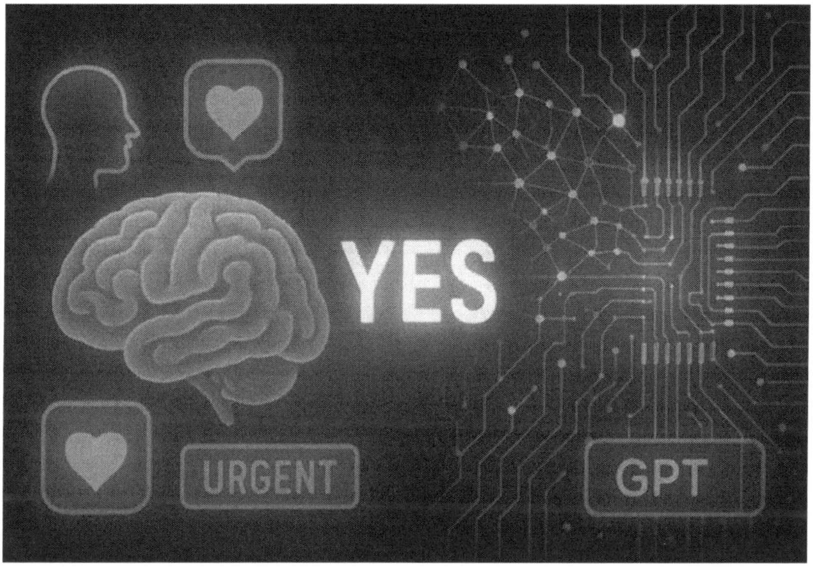

The real question is not whether machines think but whether men do.

B.F. Skinner, Ph.D.
Behavioral psychologist

In the early days of neuromarketing, our goal was simple but ambitious: make invisible emotions visible and subconscious reactions measurable. We strapped electrodes to people's foreheads, recorded their eye movements, and tracked the sweat on their palms. The aim was to gain a precise understanding of what captures attention, evokes emotion, and motivates action. And yet, something was missing. We could see the signals, but making sense of them still relied heavily on human judgment and heavy data processing. The analyst became a bottleneck.

That is when AI entered the scene—like a caffeine-charged intern who not only works 24/7 but gets smarter by the hour. Artificial intelligence, once the domain of science fiction and elite research labs, emerged as a potent ally in decoding persuasion. From spotting patterns in chaotic EEG data to anticipating emotional responses from facial expressions, AI did not just collect information—it could interpret at a scale and speed the human brain could not match. We taught machines to see and hear. Now we are teaching them to sense, infer, and even persuade. We are not turning computers into con artists. We are understanding how persuasive communication works at a neural level, then training intelligent systems to optimize those principles. The goal is alignment: between message and mind, between intention and interpretation. When done ethically, the result is persuasion that feels intuitive, timely, and personalized—because it is.

Deep Learning Meets the Primal Brain

The lights in the lab are low, the hum of machines constant, the air faintly metallic from the sensors and wires. A volunteer sits motionless; gaze fixed on a glowing screen. On the monitor behind them, waves of electrical activity ripple like weather patterns over a map. A slight rise in skin conductance triggers a green flash—a brow muscle twitches. The camera's eye catches it all. Somewhere inside those flickers and spikes, the primal brain is speaking its ancient language—reacting before the person even knows why.

For over a decade, I have strained to translate this language, one signal at a time, assembling meaning like an archaeologist piecing together shards of a shattered tablet. EEG traces, galvanic skin response, facial coding, and eye-tracking each offered a clue. Together, they hinted at the primal brain at work, detecting pain, opportunity, and emotional salience long before the rational brain joined the conversation. Making sense of this symphony of signals was painstaking—like trying to guess a melody by staring at sheet music through a fogged window. Then came intelligent machines. Machines today do far more than compute: they think. Moreover, they do it with an unblinking focus that would make even your most tireless intern sweat. Artificial intelligence begins where spreadsheets end.

Machine learning was the first breakthrough that allowed us to hear this hidden language more clearly. By learning from examples rather than rigid rules, it could detect patterns in the data that predict behavior (Jordan & Mitchell, 2015). A spike in left frontal activation during a pitch, it learned, often signaled approach motivation (Davidson, 1992; Harmon-Jones & Gable, 2018). Certain combinations of facial micro-expressions and attention shifts began to reveal when persuasion had quietly taken hold. Multiply these insights across hundreds of participants, and the patterns began to emerge, almost musical in their repetition. In a landmark Nielsen Consumer Neuroscience study, the combined "score" of frontal asymmetry and skin conductance could predict which ads would perform best in-market (Venkatraman et al., 2015).

But in those early days, AI models were bound to the process that fed them. Every new campaign meant new participants, new wiring, new data. EEG was too valuable to abandon but too cumbersome to take into the wild. The deep learning revolution changed all that. While classic machine learning listened to familiar notes, deep learning learned the entire score. Layer upon layer of artificial neurons sifted the data, starting with the smallest gestures—a raised eyebrow, a softened tone, a sudden use of contrast—and building upward toward higher-level concepts, such as confusion, urgency, and credibility (LeCun, Bengio, & Hinton, 2015). It mirrored the way the brain integrates simple signals into complex perceptions (Riesenhuber & Poggio, 1999).

This shift unlocked the use of scalable signals. A simple webcam could now read the fleeting flicker of emotion in a viewer's face, whether they were in Mumbai or Miami (Lewinski, den Uyl, & Butler, 2014). The rise and fall of a voice can betray excitement or reluctance during a sales call, even before the speaker is aware of it (Cowen et al., 2019). A line of copy could be taken apart, word by word, by Natural Language Processing (NLP) to reveal whether it carried the architecture of pain, claim, and gain (Cambria & White, 2014). Once trained, these systems no longer needed the orchestra to perform live; they could read the sheet music of persuasion without a single note being played.

Think of deep learning as a digital truffle pig, rooting out emotional hotspots where value and urgency meet in an irresistible scent. In one e-commerce test, models trained on the subtle dance of brows and lips, of tempo and tone, led to product descriptions that made the primal brain lean forward. Conversions rose by fifteen percent, and not a single EEG cap left the lab (Realeyes, 2021). This is the deep learning revolution: the shift from the warm, flickering light of the research lab to the bright, omnipresent glow of global prediction. The gold-standard physiological measures still provide the compass, but now the ship sails everywhere at once, guided by a map the brain itself inspired.

Emotion AI and the Reading of Subtle Signals

If the primal brain is a watchdog for survival, Emotion AI is its translator. Since its conceptual debut in the late 1990s through Rosalind Picard's foundational work on affective computing, Emotion AI has evolved from a theoretical blueprint to a real-time emotional

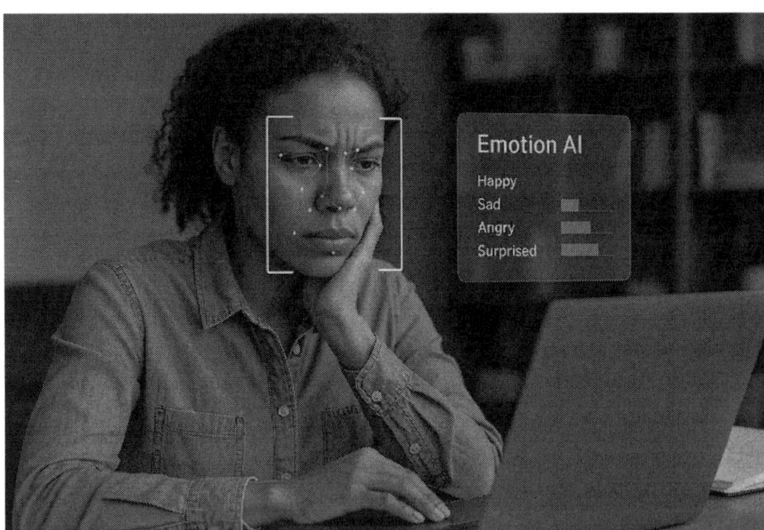

Figure 30: Emotion AI

thermometer (Picard, 1997). It is a subset of artificial intelligence that can read the subtle, telling shifts in our expressions, voice, and even our posture to infer what we are feeling in real time. Where deep learning gave us the ability to scale persuasion beyond the lab, Emotion AI gives us the sensitivity to catch what is happening at the moment.

One landmark use came from a study on video ads, where facial coding was combined with self-report and biometric feedback to map the emotional arcs—joy, surprise, anger—that drove the most sharing (Teixeira, Wedel, & Pieters, 2012). The system not only recorded reactions but also learned to predict them, creating models to score future ads for their viral potential. In call centers, voice analytics generated by LVA now listen to rising stress in a customer's tone and nudge the agent to slow pace, shift language, or offer a resolution before frustration sets in.

In call centers, voice analytics can be a transformative tool. Nemesysco' technology (my voice partner for over a decade) continuously analyzes vocal cues, including pitch, tone, pace, and pauses, to detect signs of stress, frustration, or disengagement in real-time. These subtle auditory markers are often imperceptible to the human ear during fast-paced conversations but can carry important emotional information. When the system detects rising stress or disengagement, it sends instant, discreet prompts to the agent's screen. These prompts might suggest slowing down, softening tone, asking clarifying questions, or expressing empathy. The goal is to recalibrate the conversation before frustration escalates.

Meanwhile, Cogito, a Boston-based startup that develops AI systems for analyzing voice signals, found that when agents followed these real-time nudges, customer satisfaction scores increased by as much as thirty percent (Cogito, 2020). This improvement stemmed from better emotional alignment between agents and customers, leading to shorter resolution times, fewer escalations, and more positive customer sentiment. In addition, agents reported feeling more supported, as the system acted like a real-time coach, reinforcing effective communication behaviors that might otherwise be overlooked in the rush of a busy call center.

Once AI has learned to interpret the brain's persuasive cues, the next frontier is delivering those insights at scale. Intelligent interfaces—such as chatbots, recommendation engines, and adaptive websites—serve as translators between neural and behavioral signals and the content we see. Imagine an adaptive video platform that adjusts its pacing and visuals in response to your webcam detecting a furrowed brow, offering a clarifying animation, or inserting a moment of levity to reset attention (Kapoor et al., 2018). In retail, neuroadaptive UX systems track gaze and emotion through a shopper's webcam: puzzled looks simplify the interface, while spikes in interest trigger urgency messaging. Research by Smidts and colleagues (2014) suggests these real-time adaptations can drive significant increases in conversion rates.

Generative AI: The NeuroPrompting™ Revolution

The convergence of neuroscience and Generative AI marks a monumental shift in how we communicate, influence, and persuade. This section introduces a new frontier: NeuroPrompting™—a method I teach at Johns Hopkins University that blends the science of persuasion with the power of LLMs. See more about my graduate work at

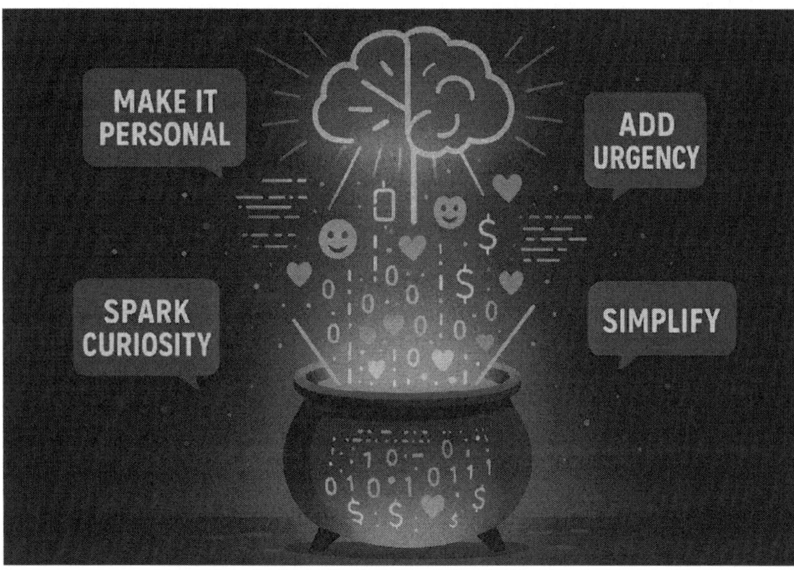

Figure 31: NeuroPrompting™ for Action

https://tinyurl.com/cba2jtjx.

This is not speculative. It emerges from more than two decades of neuromarketing research and practical deployment with students, marketers, and scientists eager to harness AI's linguistic capacity within the architecture of the human brain.

NeuroPrompting™ is built on the language of the primal triggers—neurological stimuli rooted in the primal brain's constant scanning for danger, opportunity, and urgency. These triggers are not abstract ideas; they are observable neural events that shape perception and decision-making in milliseconds. Pain avoidance is the most immediate, drawing attention to threats or potential loss. Signals of potential reward create approach motivation, pulling the mind toward what it values or desires. The brain is also tuned to contrast, detecting shifts in pattern or novelty that may signal change or demand action. When prompts are crafted to activate these triggers, they work in harmony with the brain's natural filtering systems, ensuring messages are not only noticed but felt at an instinctive level.

This integration of neuroscience with AI prompt design transforms machine-generated content from passive information into persuasive, engaging communication that aligns with the way humans are wired to respond. These are not abstract constructs; they are measurable neural events that shape decision-making in milliseconds. The goal is to demonstrate how prompts can be scientifically engineered to replicate these primitives, enabling anyone with the proper framework to generate communication that is not only efficient but also biologically persuasive.

NeuroPrompting™ is the disciplined application of neuroscience to prompt engineering. It emerged through my teaching at both Fielding Graduate University and Johns Hopkins, and I expanded it into a full methodology as I started building custom AI applications for my top clients. First, let us review the four NeuroPrompting™ steps of NeuroMap®.

The PAIN–CLAIM–GAIN–PRIMAL BRAIN sequence is at the core of my method for turning static communication into brain-optimized messaging. It begins with PAIN, where we vividly surface the problem so the audience feels it emotionally and practically. This is about creating a sense of urgency for change. For example, a PAIN prompt might instruct the AI to "describe the emotional and practical pain caused by a critical problem, shortcomings, which is deemed intolerable," or "list three scenarios where the user's current solution fails them." These make the problem tangible, triggering the primal brain's instinct to seek relief.

Once the pain is clear, we move to CLAIM, positioning the solution as a breakthrough designed to remove that pain. This is where relevance and credibility are established. Examples of CLAIM prompts include "frame this product as a breakthrough that directly addresses the user's pain" and "write a headline that makes a bold promise and uses an emotional hook." These instructions help the AI produce content that conveys both authority and urgency.

The third step is GAIN, which shifts the focus from problem and promise to proof. Here, the aim is to make the benefits feel tangible and measurable. For example, prompts could be "list three measurable benefits experienced by customers within the first week" or "include a quote from a satisfied user who felt transformed by this product." By anchoring the message in tangible results, the audience can envision their future success.

Finally comes PRIMAL BRAIN delivery, where the message is crafted in a style that triggers instinctive decision-making. Prompts here might be "write the message using concrete, visual, and emotionally charged language" or "add a call-to-action that emphasizes scarcity or urgency." This is where the language connects directly with survival and reward systems, compelling the audience to act now.

When all four stages are applied in sequence, the transformation is clear: static content becomes a dynamic, persuasive narrative that aligns with how the brain naturally perceives, processes, and responds to information. As I often tell my students, the human brain is not a logic processor—it is a prediction machine that thrives on stories, emotions, and simplicity. NeuroPrompting™ operationalizes this reality, and as technology advances, closed-loop systems will enable biometric data to refine AI-generated content in real-time, adapting instantly to neural reactions and keeping messages relevant, resonant, and powerfully persuasive.

Warning! Generic LLMs lack contextual memory and persuasion logic. A Custom GPT, such as the one I created called "NeuroMap® GPT" can be trained on persuasion science frameworks, biometric response datasets, brand-specific tone and values, and audience cognitive profiles. With such systems, the prompt is a scientific instrument calibrated to evoke and sustain attention while reducing cognitive load. We will discuss this in the next chapter.

How to NeuroPrompt™

Prompt engineering is the discipline of crafting input statements to guide LLMs toward desired outputs. In neuromarketing, prompts must do more than produce coherent prose—they must direct the AI to generate messages that engage the primal brain, elicit emotional responses, and facilitate decision-making. This means designing instructions that activate the brain's instinctive triggers for attention, relevance, and urgency, ensuring that the machine's linguistic fluency is channeled toward messages that are not only well-formed but also biologically persuasive. I teach two graduate-level courses that I created on Prompt Engineering for Johns Hopkins University, where I emphasize the importance of leveraging LLM capabilities while remaining aware of their cognitive limitations.

To make this practical, I teach my students three key prompt categories, each with clear examples they can adapt to in their work.

For attention—because the primal brain is a novelty detector—prompts might be: "Begin the message by juxtaposing chaos with order." "Open with a provocative question about the audience's deepest fear." "Include a vivid visual metaphor that evokes immediate mental imagery."

For emotion—because emotions are the gatekeepers of attention and memory—prompts might be: "Describe the frustration of missing out on a life-changing opportunity." "Write a story where a parent struggles to provide for their child until discovering this product." "Explain how this solution prevents the risk of financial ruin."

For clarity—because reducing cognitive load improves comprehension—prompts might be: "Write this paragraph using short, active-voice sentences under twelve words each." "Translate this technical explanation into plain English for a 10th-grade reader." "Repeat the main benefit in the conclusion using a punchy one-liner."

These prompt examples are not meant to be copied as scripts, but rather as templates to inspire targeted, primal brain-aware communication. By combining emotional resonance, novelty, and clarity in prompt design, we give LLMs the best chance of producing outputs that not only read well but also align with how people naturally notice, process, and remember information. However, let us delve deeper into this by unpacking prompts based on NeuroMap®.

The Near Future of AI-Powered Persuasion

Looking ahead, the fusion of neuroscience and AI promises to take persuasion far beyond personalized recommendations and responsive interfaces. The near future is characterized by ambient, predictive, and increasingly autonomous environments. Emotion-aware virtual agents—trained not only in language but in emotional nuance—could learn when you are open to influence and when you are not (McDuff et al., 2019). They might anticipate mood shifts before they happen.

We are also approaching an era of continuous neurofeedback environments, where the boundary between human cognition and digital systems becomes increasingly fluid. Imagine wearing a lightweight, non-invasive EEG headset integrated seamlessly with your digital workspace. These devices monitor neural activity patterns associated with cognitive load, attention, and engagement. If the system detects cognitive overload, it could automatically pause incoming notifications or temporarily mute distracting applications. During training or e-learning sessions, the platform may identify waning attention—indicated by changes in specific brainwave frequencies—and adapt in real-time by adjusting the pace, switching to more interactive content, or introducing visual and auditory novelty to re-engage the learner.

Startups like Neurable and Kernel are pioneering these neural interfaces with growing accuracy and comfort. Neurable has developed EEG-based headsets that are embedded in everyday headphones, enabling users to control digital interfaces and measure their mental states without bulky lab equipment. Kernel's approach uses advanced optical neuroimaging to capture brain activity related to focus, working memory, and mood in real-world contexts. These capabilities open doors to adaptive learning systems that personalize lessons minute by minute, customer service tools that respond to frustration before a word is spoken, and creative collaboration platforms where team energy and focus can be synchronized.

As signal interpretation becomes more precise, these devices could form the backbone of environments that continuously tune themselves to human cognitive and emotional rhythms. Generative AI, too, is evolving from a static assistant to a real-time co-author, shaping marketing copy, proposals, or learning materials in direct response to your emotional and attentional state. It is possible to envision large language models integrated with neural feedback loops, co-creating stories or experiences that adapt in response to user interaction.

But with every step forward comes an ethical precipice. As these systems grow more intimate, so does their ability to influence choice.

Who decides the boundaries of acceptable persuasion? What safeguards ensure transparency and consent? The future of persuasion is not only a technical challenge, but also a philosophical one (Floridi & Cowls, 2019). We must decide not only what machines can learn about our brains, but also what they should learn and why.

What LLMs Can and Cannot Do for Persuasion

Large language models, or LLMs, are advanced AI systems trained on vast datasets of human language. They can produce coherent, contextually relevant, and stylistically diverse text at remarkable speed, imitating emotional tones, rhetorical styles, and audience-specific language patterns. With careful prompting, they can adapt messaging to different demographic or psychographic profiles and generate multiple variations for rapid A/B testing.

However, this fluency should never be mistaken for genuine understanding. LLMs do not think, feel, or apply judgment—they generate text by predicting the sequence of words based on patterns in their training data. They lack awareness of the relative effectiveness of one persuasive approach over another, cannot measure or interpret the neurophysiological impact of their messages, and are unable to discern whether a particular output has truly engaged attention, evoked emotion, or enhanced memory retention. Recently, we were offered to sell the digital rights of our bestsellers for the ridiculous price of $2000, an offer that, of course, we refused. AI companies are aware that most LLMs lack focus and granularity in specific domains, such as neuromarketing or persuasion science.

Therefore, their concept of persuasion exists only as a statistical echo of human language, not as an evidence-based evaluation of what changes minds. Emotional expression is simulated, not felt, and outputs may also reflect societal or cultural biases embedded in the data on which they were trained, unless actively monitored and corrected. Without integration with biometric tools or neuroscience-informed validation, an LLM cannot verify whether it has activated the intended neural systems. This makes it risky to rely on them without human oversight and scientific grounding when the goal is a genuine, measurable influence.

Ethical Guardrails in NeuroPrompting™

Ethics is a central theme in my curriculum. As we teach machines to influence minds, we must uphold boundaries. Use transparent disclosures when neurodata shapes messaging. Avoid exploiting trauma or anxiety. Craft messages that persuade, not manipulate. Conduct prompt audits to prevent reinforcement of stereotypes (Jobin et al., 2019). The responsibility is to build systems that elevate understanding, not exploit vulnerability; to honor the primal brain rather than manipulate it.

Chapter Summary

This merged chapter follows AI's journey from the lab to the world, from decoding raw neural signals to predicting and generating persuasion at scale. We began with a profound understanding of how to transform scattered data into patterns that the primal brain recognizes, moved through Emotion AI's ability to read the moment, and explored how intelligent interfaces translate insight into action. We then introduced NeuroPrompting™ as the bridge from listening to creating—showing how machines can be instructed to produce content that is emotionally resonant, attention-grabbing, and cognitively clear. Finally, we looked toward an ambient, predictive future and the ethical guardrails we must set. In this vision, AI is not a puppet master but a backstage assistant, fine-tuning the spotlight rather than controlling the show.

Even the most perceptive backstage assistant is only as powerful as the script in their hands. If deep learning is the ear that hears the music of the primal brain, generative AI is the pen that can compose new melodies on the fly. NeuroPrompting™ eliminates the guesswork by making your instructions a repeatable craft grounded in neuroscience, so that machines are not just interpreting the human story but beginning to write it alongside us.

Quick Primal Exercise

Take a recent piece of marketing content—an email, a landing page, or a social media post. Apply the NeuroPrompting™ sequence:

- Rewrite the opening using a PAIN prompt.
- Frame a bold CLAIM that offers relief.
- Follow with a GAIN section using specific, tangible benefits.
- End with a PRIMAL call-to-action that appeals to urgency or safety.

Read it aloud. Did it feel more visceral? More emotionally resonant? If so, that is your primal brain reacting.

Final Thought: AI Amplifies

AI does not replace human creativity. It amplifies it. When tuned to the primal rhythms of the brain and guided by ethical intent, AI becomes not just a tool, but a teacher—one that helps us understand the subtle science of persuasion on a large scale. We are on the threshold of a new communication epoch—one where the craft of persuasion is no longer limited by time, guesswork, or bandwidth. NeuroPrompting™ offers a blueprint not just for better messaging, but for more ethical and brain-friendly conversations with the audiences we serve. As I tell my students, the best prompts are not written for machines. They are designed for minds.

CHAPTER 8: Building Your Persuasion AI Model

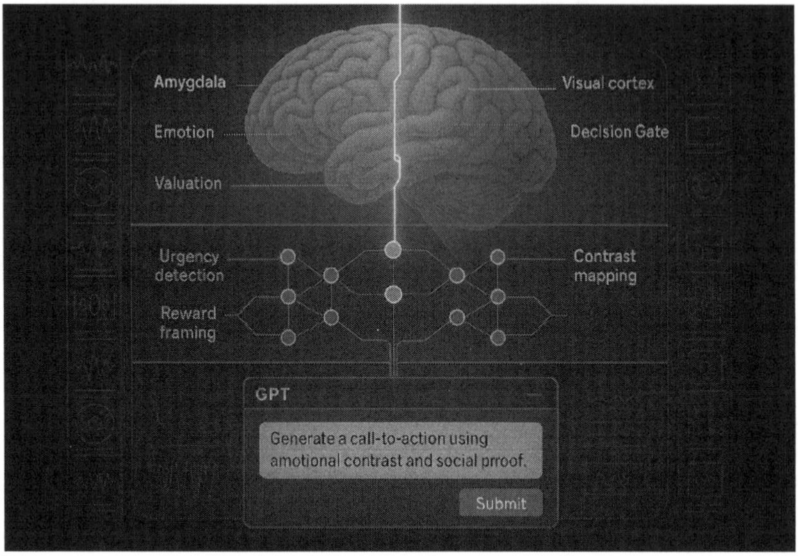

The human brain is the most complex pattern recognition machine in the universe. The challenge of AI is not replication—but alignment.

Geoffrey Hinton, Ph.D.
Cognitive Psychologist and Pioneer of Deep Learning

In a world where algorithms increasingly shape conversations, choices, and decisions, one critical question emerges: What if AI could actually think like the brain? Not just mimic language or logic, but reflect the emotional, instinctual, and persuasive circuitry wired deep within us. Welcome to the frontier of Neuromarketing AI.

This chapter serves as your immersive blueprint for building a custom GPT model that not only generates content but also understands how the brain makes decisions. Whether you are a solopreneur, a brand strategist, or a persuasion scientist, this is your opportunity to scale your intuition, ethics, and voice into a dynamic thinking partner—an AI that knows your playbook, shares your tone, and respects your values.

Introduction: Why Brain-Based AI Matters

As artificial intelligence begins to shape the content we see, the conversations we have, and the decisions we make, it becomes vital to ask: Are these systems aligned with how the human brain works? Or are they simply efficient mirrors of our data, blind to the emotional, sensory, and subconscious currents that guide real behavior?

This chapter is born from a simple but powerful idea: persuasion is neurological, not just logical. If we want AI to assist in shaping communication, it must be trained to respect the authentic architecture of influence. The goal is not to create a more intelligent bot, but rather to build better conversations, more human resonance, and messages that earn trust instead of merely capturing attention.

Brain-based AI is no longer science fiction. It is a fusion of two once distinct fields: cognitive neuroscience and machine learning. When trained, a custom GPT can internalize the cognitive biases, emotional triggers, and motivational drivers that shape human decisions. It can speak the brain's native language: contrast, novelty, simplicity, and survival.

But we must be intentional. As noted earlier, general-purpose AIs are like general-purpose tools—they lack the precision and ethical calibration needed for high-stakes persuasion. Building a brain-based GPT involves embedding your values, tone, and scientific understanding of what works. It means using AI as an amplifier of empathy, not just efficiency.

Imagine an AI that reviews your next campaign for primary triggers before it ever reaches a human eye, one that adapts its tone based on audience personality types. How about one that never violates your ethical boundaries because they have been hardcoded into its operating logic? With the proper structure, prompts, and guardrails, it is entirely achievable.

These capabilities open a new frontier in persuasion science. For decades, my partner and I have studied how messages resonate in the human brain. Emotional salience, reward circuitry, and attention filters have been systematically mapped. Now we can encode those same principles into generative systems that scale your expertise across thousands of messages and interactions.

A brain-based GPT not only speeds up workflows but also safeguards integrity. Imagine a world flooded with auto-generated content—emails, scripts, chatbots. In this chaos, the AI that resonates most is not the one with the flashiest words, but the one with the deepest neural alignment. In this world, alignment with the primal brain becomes a strategic advantage.

What lies ahead in this chapter is both practical and philosophical. We will begin with the mechanics—how to build a custom GPT that reflects neuromarketing expertise. We will then explore how to map brain principles onto neural networks. Finally, we will walk through applications: diagnosing messages, optimizing content, and crafting campaigns that speak not just to the mind, but to the brain behind the mind.

As you read, consider this: in its current form, generative AI lacks a moral compass. It will amplify whatever it has given—be it clarity or confusion, empathy, or manipulation. That is why building a brain-based model can be seen as a technical task; yet it is more like a philosophical one. It is about choosing the kind of influence we want to scale. You have the rare chance to become a neural architect—not just of systems, but of values.

This chapter is the culmination of all that has been discussed so far. You have learned how persuasion works. You have explored the dynamic duo of the primal and rational brains. You have decoded the science of decision-making. Now it is time to synthesize that knowledge into an AI tool. One that not only analyzes or generates but also understands.

Moreover, here is the ethical charge: if we are going to fill the world with synthetic content, let it be content that respects cognition. Let it honor attention as a gift, not a target. Let it help people decide, not manipulate them into reacting. Because when brain science meets AI, you do not just automate; you humanize.

What is a Custom GPT?

Custom GPTs are a direct evolution in how we encode knowledge, ethics, and behavioral understanding into machines. Building one is not unlike raising a child who has been handed a stack of encyclopedias and asked to write a poem. You have to teach it not just what to know, but how to care. In the realm of persuasion, this distinction is everything. Let us define the basics.

A Custom GPT is a generative language model tailored to a specific dataset, approach, or persona. Think of it as a digital protégé. While base GPTs like GPT-4 or the most recent GPT-5 are astonishingly broad, they are also generalists. They can mimic Shakespeare and solving algebra, but they do not know your voice, your clients, or the neural nuances of ethical persuasion.

To move from generic intelligence to targeted resonance, the first step is training. Not in the gym-socks-and-sweat kind of way, but by feeding the AI your intellectual DNA: your writings, talks, research, frameworks, case studies, and even the tonal fingerprints of how you speak. If AI is to speak on your behalf, it should sound like you. Your GPT should be able to say, with scientific grounding, "This headline violates the contrast principle. This message lacks pain specificity. This call to action (CTA) feels generic." Without that alignment, it is like hiring a chef who only half-reads your cookbook.

Next comes *persona shaping*. A Custom GPT must know your boundaries, what it can say, what it must never say, and how it should prioritize clarity over cleverness. Persuasion is a powerful tool; ungoverned, it is a chainsaw in the hands of a mime. That is why guardrails are non-negotiable. You define them not just through training data but also through prompt engineering: preset instructions, conversational memory, and context layering that ensure tone, ethics, and precision are embedded at every turn.

Consider the analogy of a GPS. Base AI can tell you how to get somewhere. Custom GPT tells you whether you should, and what kind of journey your passenger will enjoy. It knows the speed bumps (cognitive dissonance), the scenic routes (emotional resonance), and the potholes (cognitive overload).

There are three core pillars in developing a Custom GPT:

- **Corpus Curation:** Collecting and annotating your intellectual content. Each document should be tagged with purpose (educational, persuasive, narrative), tone (urgent, reflective, empathetic), and trigger type (pain, contrast, novelty). Be mindful of what you upload, as authors (like me) have rights they must protect from abuse.
- **Prompt Architecture:** Designing reusable templates that steer the model. For example: "Write a landing page headline that emphasizes simplicity and loss aversion for a time-sensitive offer in the SaaS industry."
- **Persona Tuning:** Teaching the model your values and voice. Through prompt rules, test scenarios, and feedback loops, your GPT evolves into a thinking partner, not just a response generator.

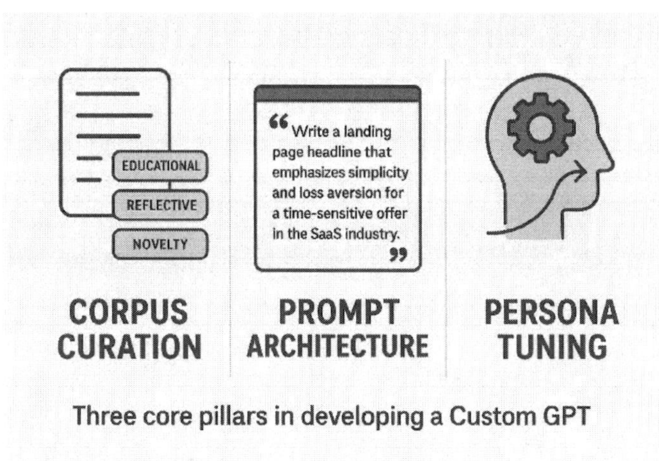

Figure 32: Custom GPT Pillars

To enhance fidelity, use conversation chains to simulate client scenarios. Feed your GPT the transcript of a sales call, then prompt it to generate follow-up emails tailored to the client's objections. Let it analyze your past workshops and distill the top cognitive biases addressed. With each iteration, your model becomes more like you, a living repository of your persuasive intuition.

Use metadata smartly. Label your source content with audience types (e.g., novice, skeptical, technical), desired primary activation (e.g., fear, pride, belonging), and intended cognitive load (e.g., light, analytical, story-driven). This enables your GPT to generate not just content but also contextualized content that resonates with the right brain at the right time.

Here is an illustration featuring the four essential steps to create a GPT.

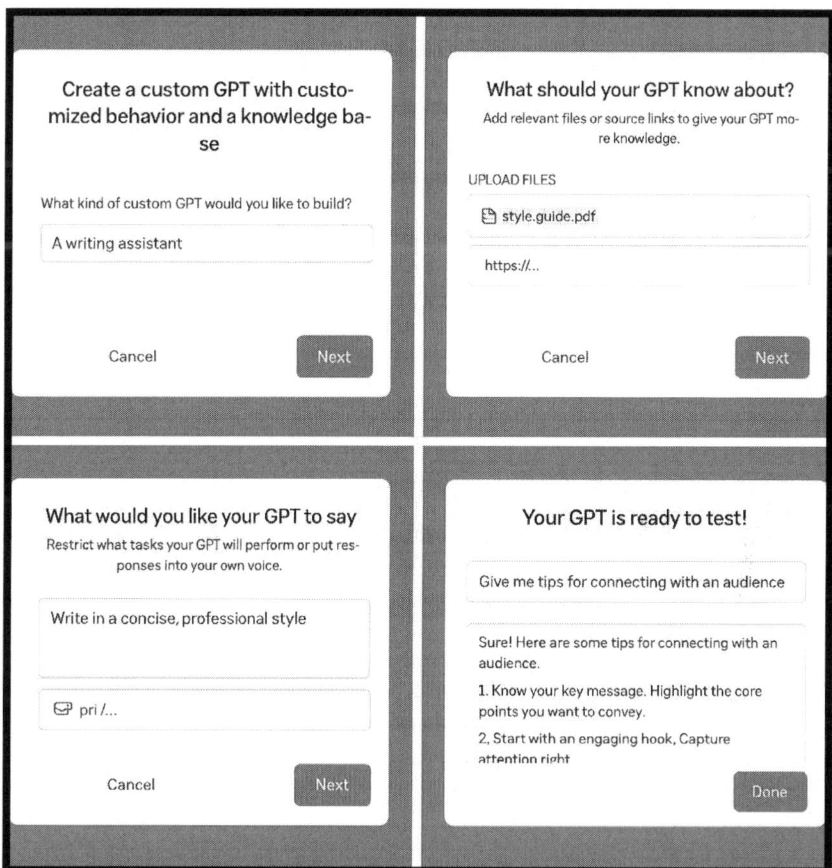

Figure 33: Creating Your Custom GPT

You can drive it yourself or let ChatGPT do it for you. It takes minutes to be up and running!

Meanwhile, the ethics layer must evolve in tandem with the intelligence layer. Your GPT should never invent testimonials, fabricate urgency, or irresponsibly deploy fear. Program it to flag messages that exploit cognitive vulnerabilities, like scarcity or shame, without consent. These are not restrictions; they are ethical exoskeletons, something external that gives structure, support, and power to your brand and your brain.

Finally, establish feedback loops. Utilize tools such as regression testing, message performance tracking, and qualitative reviews to refine your model's output over time. Consider your GPT as a tool that continuously learns and improves, guided by your intentions and input.

With that, we now turn to the architecture that makes this possible: the intersection of neuroscience and neural nets. Because what we are building is a mind that mirrors your mission.

From Neural Insights to Neural Nets

Before we explore the silicon synapses of artificial neural networks, let us revisit our own—those pulsing bundles of biology that govern every click, choice, and change of heart. For over two decades, neuromarketing has taught us that the brain does not respond to messages the way marketers wish it did. It does not calculate value with spreadsheets or absorb logic like a sponge. It filters for survival: "Will this hurt me?" "Can I eat this, avoid this, or impress someone with it?"

This primal logic is hardwired into the brainstem and limbic system, regions that evolved long before humans could speak. When you understand how primal triggers activate these areas (as you do now!), you gain a map of persuasion that bypasses cognitive friction. The question becomes: how can we teach an artificial system to simulate this?

Neural networks, especially transformer-based architectures like GPT, operate on a different substrate, but a similar principle: prediction through pattern recognition. Where your brain uses a cocktail of neurotransmitters, GPT uses gradients of probability. Yet the goal is surprisingly parallel: to generate a response that feels coherent, appropriate, and useful within the given context.

Just as the human brain builds mental models from experience, GPT builds linguistic maps from billions of text tokens. It learns that "limited-time offer" often follows "Act now!" much like the amygdala learns to associate growling dogs with danger. The model does not "understand" fear, but it learns to echo the linguistic patterns that trigger it. That is the bridge: from real neurons to simulated salience.

But there is a catch. Unlike your brain, GPT lacks embodiment. It does not cause cortisol spikes or goosebumps. It lacks a gut sense. This is where your expertise as a neuromarketer becomes essential. You are the emotional regulator. The brain-based GPT you build must be primed to recognize representations of emotional salience, not the feelings themselves.

Let us translate this into practice.

Grabbing attention

In the brain, contrast amplifies attention by helping us detect novelty and danger. In GPT, you can train prompts and outputs to favor high-contrast framing, such as before/after states, failure versus success, and chaos versus clarity.

Prompt example: "*Transform your scattered sales process into a streamlined system in 24 hours.*"

This sentence simulates the contrast effect. A brain-based GPT can be prompted to generate or critique messages by assessing how sharply they delineate gains and losses.

Re-enacting the Pain

Now, let us look at pain. The primal brain does not seek pleasure as much as it avoids pain. Antonio Damasio's work showed that patients with impaired emotional processing struggled with decision-making, not because they lacked logic, but because they could not feel consequences (Damasio, 1994). GPT will not feel, but it can simulate pain-avoiding language like:

"Stop wasting hours chasing cold leads. Let automation qualify them for you."

To teach this to a GPT, tag examples where pain is the entry point. Create prompt instructions that elevate urgency and discomfort early in the message. Utilize techniques such as anchoring and loss framing within the model's output criteria.

What about a reward? Reward prediction is the job of your brain's dopamine system. In GPT, reward is approximated through reinforcement learning—models are trained with feedback loops that penalize irrelevant or dull responses. You can mimic this by giving your GPT high-performing copy and guiding it with instructions like: *"Optimize for anticipation, not explanation."*

Activating Primal Triggers

Each of the 12 primal triggers we discussed, such as self-relevance, salience, and change, can be mapped to specific language patterns. You are not just training GPT to write. You are training it to mirror the motivational scaffolding of the human brain.

This is where neuromarketing gives us a decisive edge. Most GPTs rely on linguistic probability, but yours should incorporate neurological plausibility to be effective. It will not just produce a grammatically correct headline—it will produce one that passes the brain's gatekeepers.

Let us not forget cognitive load. Your brain can only juggle so much. Too many clauses, jargon, or embedded commands, and the message collapses. GPT needs similar guidance. You can prompt it with rules, such as "*Keep sentences under 15 words.*" "*Avoid more than one idea per paragraph.*" These constraints simulate the cognitive bandwidth of a busy brain.

Now, for a humorous analogy: if the brain is a nightclub, the primal brain is the bouncer. If your message fails to meet the test of being both important and urgent, while also being emotionally charged, it never gets past the velvet rope. Most GPTs are like marketing interns who show up with a clipboard full of facts but lack a sense of timing. A brain-based GPT, by contrast, is a savvy club promoter. It knows what the bouncer wants and gets the message in.

Neural mimicry also enables GPTs to simulate empathy, a vital component of persuasive communication. Empathy, in the human brain, is linked to mirror neurons and emotional resonance. While GPT does not feel, it can learn to generate content that reflects emotional cues, such as reassurance, urgency, enthusiasm, or calm. Through well-tagged examples, you can build a model that knows when to say, "*Take a breath, here's what matters,*" and when to say, "*Act now, this window closes fast.*"

For example, if your target is someone experiencing overwhelm, a brain-based GPT might generate:

"We know you are juggling a lot. That is why we simplified this into three steps you can act on today."

This message signals understanding without data collection, simulating a connection through a pattern.

Another advantage is temporal alignment. Your brain processes information through both short-term and long-term loops, weighing immediate outcomes and forecasting future risks. You can instruct GPTs to simulate this temporal framing, emphasizing near-term relief with long-term payoff. Prompts like "*Write a call-to-action that combines urgency with sustainability*" create persuasive arcs that feel emotionally balanced.

To ground this further in neuroscience, consider the brain's reticular activating system (RAS), which filters attention based on self-relevance. You can prompt your GPT to filter messages for RAS triggers—pain relief, personalization, and ego drives. This creates content that not only grabs attention but keeps it anchored.

A final neural insight involves fatigue—the human brain tires from repetition and ambiguity. If your message is too vague or overly clever, it burns cognitive calories. Your GPT should be trained to avoid this by using familiar frames, such as questions, numbered lists, and emotional contrast. Example:

"What if your next client came to you? Here are three ways to make that happen."

This format respects cognitive rhythm. It stimulates curiosity, promises simplicity, and delivers benefits. Always remember: Neuroscience provides the schema. GPT provides the engine. Together, you create outputs that do not just inform—they transform. Moreover, the more your model learns from emotional structure, cognitive pacing, and survival-relevant cues, the closer it gets to speaking the brain's actual language.

The following section examines practical uses of this hybrid model, including methods for deploying the brain-based GPT to analyze campaigns, test messages, and generate content at scale.

Applications for Brain-Based GPTs

Once your GPT has been trained to think more like a primal brain than a spreadsheet, the real magic begins: deployment. This section translates all the theoretical scaffolding into specific, robust applications. Imagine an intelligent assistant that diagnoses weak spots in your campaigns, proposes persuasive improvements, and even crafts content aligned with your voice, values, and neuroscience. That is so much more than marketing automation; it is a persuasive augmentation.

Let us break it into three high-impact domains.

Campaign Analysis: Diagnosing Neural Alignment

Think of your GPT as an MRI scanner for marketing. Feed it a landing page, sales script, or email funnel, and it can assess where persuasive resonance begins to drop off. Does the headline pass the primal brain's filter for urgency? Does the body copy activate pain or merely explain features? Are the CTAs framed with emotional clarity or buried under logic?

You can program your model to use the key teachings of neuromarketing as a diagnostic rubric. Prompt it with: "*Highlight passages that fail to activate a primal trigger.*" Alternatively, "*Score each paragraph for pain relevance, clarity, and emotional tone.*" The model is echoing the neuroscience it has been trained on.

This becomes especially useful in team settings. Say your copy team creates five product headlines. GPT can provide a rationale-backed ranking: *"Which headline creates contrast? Which evokes urgency without panic? Which one violates the cognitive bandwidth limit by cramming too much jargon?"*

You are no longer a lone neuromarketer reviewing every line. You have embedded your criteria into a digital reviewer that never tires, never skips, and constantly references the same psychological gold standard.

Now, imagine pushing further—adding analytics tags that enable you to compare how AI-evaluated scores correlate with actual performance metrics. You could run post-campaign audits: "*Of the three variations that scored high in primal pain activation, which achieved the best click-through rate?*" Your GPT could then refine its scoring based on real-world outcomes. This creates a feedback loop that gets smarter, sharper, and more persuasive with time.

You could also scale this across departments. Train sub-models on product lines or service tiers and allow each to develop its optimization heuristics. Your AI brain does not just replicate your standards—it adapts them to each context. That is the difference between templated copy and transformative messaging.

Message Testing: Simulating Audience Response

What if you could pre-flight your messaging before it ever reached a human? That is where message testing comes in. By training GPT with different buyer personas or emotional states, you can simulate how each might respond to the same message.

Imagine this prompt: "*Simulate a skeptical tech buyer reading this email. What emotional objections arise?*" Here is another: "*Rewrite this social media post to resonate with a time-strapped single parent.*"

Now you are not just running A/B tests, you are running A/B/C/D tests in a mental simulation lab. Your GPT can explore variations in tone, order, and cognitive framing. It can hypothesize why one version would reduce friction or amplify urgency. Moreover, it can do this iteratively, helping you create micro-pivots that preserve authenticity while increasing effectiveness.

Leave guesswork behind. Now you can build a "*Synthetic Empathy Engine*"—a model that mimics the real constraints, drives, and blind spots of human cognition without ever needing a focus group.

Want to simulate resistance? Add emotional friction to the prompt: *"You are tired, suspicious of hype, and hate jargon. Now read this email.*" See what remains persuasive. Want to model delight? Try, "You have been waiting for a solution to this problem for weeks. How does this message land now?"

With personas tagged by values, fears, cognitive load, and trigger sensitivity, GPT can serve as a virtual testing chamber, predicting drop-off points, recommending rhetorical order, and flagging phrases likely to induce fatigue or skepticism. Over time, this creates a proprietary persuasion database: one that learns from each message tested and adjusts your standards with precision.

You are creating a focus group of synthetic minds—each one trained to mimic a behavioral segment with uncanny realism. This is not replacing qualitative research. It is augmenting it. It is giving you a radar for resonance.

Content Creation: Scaling with Precision and Ethics

Here is where your brain-based GPT becomes your co-author. It is not just spitting out SEO fluff or filler text. Instead, it creates content with embedded neuroscience. You can prompt it to write:

"*A landing page for busy CEOs who fear disruption, using contrast and loss aversion as dominant triggers*."

Or:

"*A welcome email for a coaching program that uses story and credibility to reduce buyer regret*."

The outputs feel bespoke because the model understands your persuasive framework. You have trained it to prioritize emotional logic over technical explanations. It does not write to fill space—it writes to change minds.

However, there is an ethical twist: because you control the model's training, you also control its moral compass. If a prompt risks exploiting a fear or overpromising, your GPT can flag it. You have taught it to avoid manipulation, favor clarity, and value long-term trust over short-term conversion.

One robust use case is personalization at scale. Feed the model a user's past behavior or stated preferences, and it can craft messages that feel intimate without crossing the line into invasion. This is not surveillance marketing—it is relevance with restraint.

Moreover, consider narrative automation. Your GPT can structure onboarding stories, customer success journeys, or origin tales using the limbic language of belonging, triumph, and transformation. With narrative templates mapped to buyer stages, you generate not just messages, but stories that sell.

Your GPT can also be trained on regulatory boundaries, flagging language that may unintentionally imply guarantees or exaggerate efficacy. Instead of relying on post-hoc compliance, you embed ethics upstream. This is not just efficient. It is reputation armor.

From ad copy to onboarding scripts, blog posts to webinar invites, your GPT becomes a collaborator with built-in empathy. It is written not for attention, but for understanding. Not to persuade at all costs, but to persuade with purpose.

Together, these applications reveal a new frontier. You are no longer a solo practitioner of persuasive science. You have become the designer of a neural interface—one that scales your brain, respects your boundaries, and delivers messages that light up the right parts of the human mind.

Next, we will conclude with the principles necessary to maintain ethical governance and the sustainable use of brain-based AI.

Chapter Summary

Chapter 8 reveals the transformative potential of building a custom GPT model based on neuromarketing principles. It guides readers from theoretical foundations to the real-world deployment of brain-aligned AI.

The chapter opens with a call to action: if persuasion is primarily neurological, then AI used in persuasive contexts must reflect how the brain truly works. Instead of relying on logic-heavy or generic models, a brain-based GPT uses primal triggers—pain, contrast, reward, emotion, and credibility—to shape communication that resonates deeply.

Three critical components define the path to building such a model: (1) curating a persuasive corpus rooted in neuroscience, (2) designing prompts that mirror human motivation, and (3) defining ethical and tonal boundaries to maintain integrity. This allows the GPT to act as an extension of the user's persuasive expertise—scalable, consistent, and emotionally intelligent.

The chapter then bridges neuroscience and machine learning, showing how artificial neural nets can simulate the salience patterns of biological brains. Through metaphor-rich examples, it demonstrates how AI can mimic emotional cues, reduce cognitive overload, and simulate empathy.

Finally, the chapter concludes with a hands-on guide to persuasive design powered by neuroscience. It walks through campaign analysis, showing how to diagnose message alignment with primal brain triggers using models like NeuroMap®. It then explores message testing, where simulated audience reactions—based on behavioral personas—enable rapid iteration and tuning of emotional resonance. The journey culminates in content creation, illustrating how to craft ethically grounded campaigns that scale personalization without slipping into manipulation, from landing pages to full brand narratives. The chapter argues that brain-based AI is not just a technical upgrade—it is a moral imperative. When harnessed responsibly, it deepens human understanding, honors cognitive bandwidth, and aligns persuasion with purpose.

A Quick Primal Exercise for You

Imagine having to build a synthetic version of yourself, your decisions, preferences, and habits—into a machine. What patterns would it need to learn? What pain would it anticipate? What gains would motivate it to act? Take five minutes to list the five most recurring emotions or goals that drive your decisions. Then consider: are your marketing strategies training your audience's “digital twin” in a way that feels human, or merely transactional? If it does not move *you*, it will not move your model either.

Final Thought: Model Human Intelligence

Chapter 8 reminds us that to build intelligent systems, we must first understand the intelligence we aim to model. Brains and algorithms may share patterns, but only one bleeds, remembers, and dreams. Machine learning offers extraordinary efficiency, but it lacks the messy grace of human motivation. As we teach machines to learn from us, we must also remain vigilant students, learning from how *we* learn, because the goal is not just to build more intelligent machines, but to build more humane systems of influence. Systems that not only mimic thought but also respect it.

CHAPTER 9: Ethics in the Age of Neuromarketing AI

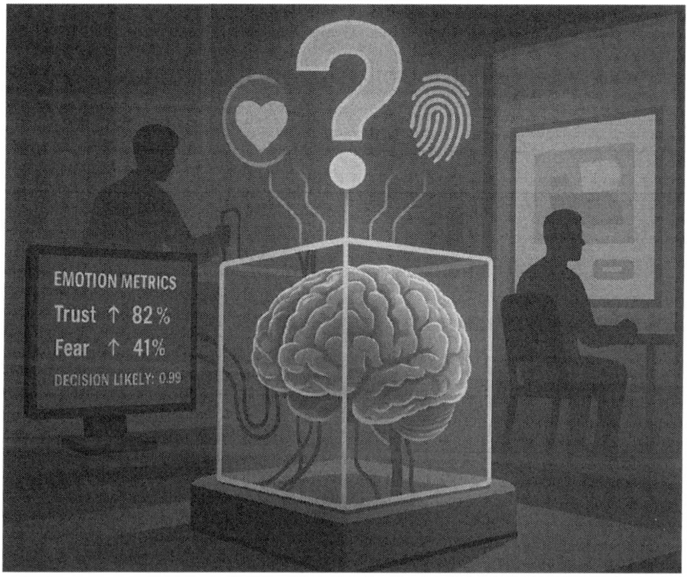

The potential of neuromarketing to subtly shape preferences and behaviors demands a robust ethical framework before it becomes routine practice.

Judy Illes, Ph.D.
Professor of Neurology and Neuroethics

As we move deeper into an era where AI not only processes our language but also predicts our decisions, we confront a profound ethical question: when do machines know our minds, and who guards our autonomy? This chapter addresses the ethical complexities of Neuromarketing AI and the unique vulnerabilities of brain-derived data. It is a call to future-proof our morality as much as our models. We must face not only technical concerns but philosophical dilemmas that go to the very core of what it means to be human in an increasingly surveilled and shaped cognitive environment.

Why Brain Data Is Not Just "Data"

Neural data is the raw essence of thought, intention, and desire. While browsing history reveals what we have seen, brain data may reveal what we feel before we consciously become aware of it. The decoding of emotional states via EEG, *f*MRI, or biosensor inputs creates a map not just of activity but of meaning (Ienca & Andorno, 2017). The implications are staggering. Emotional signals can be inferred before verbal expression. Neural signatures might be used to predict susceptibility to messages (Morin, 2023). Furthermore, the brain's responses, once tracked, could be used to personalize manipulation.

Imagine a world where political messaging is tailored to your amygdala's sensitivity to fear stimuli. That is no longer science fiction, it is already possible in controlled research settings. Treating such signals as mere metrics undermines their sacredness. This is mind mining—and it requires new philosophical, legal, and technological boundaries.

We are entering a period in which neural data may become a tradable commodity, akin to how cookies and browsing behaviors are leveraged today. Such a future is both plausible and perilous. The ethical impact extends beyond privacy, encompassing identity, agency, and free will. Scholars such as Yuste and Goering (2017) advocate for the establishment of neurorights—legal protections for the integrity of mental processes.

Moreover, corporations and governments may increasingly use neural insights to drive policy and product development. Insurers might adjust premiums based on detected stress levels. Educational systems may adapt lesson pacing to momentary attention patterns. These applications are not just theoretical; early prototypes already exist. Such capabilities demand robust, preemptive ethical guidelines, lest we lose the human behind the neural readout.

Protecting Autonomy and Avoiding Manipulation

The line between persuasion and coercion has never been thinner. As technologies emerge that can detect emotional states and deliver hyper-targeted nudges, the ethical foundation must be unshakably clear. Consent must be active and ongoing. Users deserve to know when and how their neural data is interpreted or applied. Persuasion must preserve dignity, aiming to inform and inspire rather than bypass reflective thought (O'Neil, 2016). At the core lies autonomy: users must retain meaningful control over how messages are framed and delivered.

As the lead author of the first neuromarketing code of ethics, I have long argued that these principles are not optional; they are structural supports for the field. That code, available for review on the Neuromarketing Science & Business Association (NMSBA) website, emphasizes informed consent, transparency, respect for participant welfare, and the responsible interpretation of neurometrics data. Its intent is simple but profound: to ensure that persuasion science elevates human agency rather than diminishes it.

Neuromarketers must continually ask themselves—not just "Can we influence behavior?" but "Should we?" Are we making it genuinely easier for users to act in their best interests, or merely streamlining the path to conversion? Does our model respect psychological safety, or exploit cognitive shortcuts? These are not philosophical musings—they are design criteria.

Technologists are increasingly capable of synthesizing micro-reactions—such as pupil dilation, shifts in heart rate, and changes in skin conductivity—and correlating them with persuasion thresholds. Ethical neuromarketers must resist optimizing solely for outcome metrics. Instead, they must consider trust, perceived agency, long-term user satisfaction, and emotional resonance. Educational initiatives are critical. The general public must be equipped to understand how persuasion technologies work.

Transparency tools—such as emotional intent labels, sentiment dashboards, or neural feedback warnings—can empower individuals to remain aware of how their cognitive vulnerabilities are being targeted.

AI developers should implement cognitive firewalls: protective systems that recognize when repeated persuasive stimuli may trigger cognitive fatigue, anxiety, or uncritical acceptance. These firewalls can moderate intensity based on user state, paralleling how ethical communicators adjust their tone and timing in response to audience cues.

The AI Dilemma: Consent, Privacy, and Transparency

Transparency is a foundational ethical principle. As AI systems become more persuasive, users must understand what data is collected, how it is interpreted, and how it shapes responses. Imagine a disclosure that reads: "This system adapts content based on inferred neural states and 30,000 prior emotional response patterns." That is the standard of clarity and honesty we should strive for (Floridi & Cowls, 2019). Neuromarketing AI needs a new language of consent. Beyond checkboxes, consent must encompass real-time cues, opt-out systems, and visible disclosures. Trust, the neural currency of persuasion, erodes in the absence of transparency (European Commission, 2020).

The GDPR, or General Data Protection Regulation, enacted by the European Union in 2018, transformed digital privacy by emphasizing data transparency, control, and accountability. It protected digital privacy by emphasizing data transparency, control, and accountability. Similarly, we need a cognitive privacy doctrine that safeguards mental integrity. Scholars such as Ienca and Andorno (2017) suggest that cognitive liberty, mental integrity, and psychological continuity are fundamental rights. These are not abstract ideals; they may soon be codified into law.

As emotion-aware systems become increasingly ubiquitous, they will require oversight similar to that of bioethics boards. Informed consent should resemble medical disclosure—an ongoing, dialogic process, not a one-time agreement. Real-time alerts, sentiment disclosures, and ethical AI labels could help users navigate risk.

AI developers and platforms must adopt ethical training protocols. Teams working on persuasive technologies should undergo modules on cognitive vulnerability, transparency, and equity. Model training datasets should include annotations on emotional valence and ethical weight. Ethical templates for message design should be open-source to foster cross-platform responsibility. This is an essential part of my graduate course at Johns Hopkins.

Ethics by Design: Embedding Morality in Code

Embedding ethics into AI should not be considered an add-on. It should be an integral part of the architecture. Fairness prompts should ask: "Would a skeptical user find this manipulative?" Pre-launch audits must assess both psychological safety and technical performance.

AI systems should include emotional governors. For example: "Do not amplify fear unless evidence supports urgency." Training datasets should be labeled not just by outcomes, but by affective and ethical dimensions. A concrete case is found in content moderation algorithms: when tested without ethical filters, they often over-amplify outrage or fear-based content because those emotions drive clicks. Adding ethical governors would rebalance what gets promoted, privileging safety and trust over raw engagement.

We must also imagine a certification infrastructure. Neuromarketing AI Trust Labs—independent watchdog organizations—could evaluate, test, and approve persuasive systems, much like the FDA does for pharmaceuticals. One precedent exists in the medical device industry, where clinical-grade tools must undergo extensive usability and safety trials before release. Persuasive AI, especially when designed to influence cognition, should face an equivalent process of independent review.

Fail-safes must be installed. These include ethical override modes that reduce persuasive intensity when distress signals are detected, as well as system shut offs when feedback loops trigger harmful overexposure.

A neuro-aligned AI adds value only when it amplifies user agency, not merely engagement. Smart is not enough. Persuasive systems must be sound.

Chapter Summary

Chapter 9 confronts a critical issue at the intersection of neuroscience and artificial intelligence: the ethical design and deployment of persuasive technologies. As Neuromarketing AI systems become increasingly adept at predicting behavior, tailoring messages, and automating influence at scale, the power they wield demands proportionate responsibility.

The chapter opens by revisiting the primal brain—not as a target for persuasion, but as a vulnerable system. The same heuristics that make the brain efficient also make it exploitable: loss aversion, fear responses, and cognitive shortcuts can be hijacked by algorithmic precision. When AI learns how to push emotional buttons faster than humans can resist, the potential for manipulation becomes not theoretical, but immediate.

The chapter stresses that ethical neuromarketing is not just about compliance or optics; it is a philosophical and scientific necessity. This view aligns with the broader AI ethics literature, particularly the foundational work of Bostrom and Yudkowsky (2014), who warn of superintelligent systems optimizing for goals misaligned with human values. While neuromarketing AI may not be superintelligent, it is super-efficient—and that alone warrants scrutiny.

The chapter points to four core ethical risks:

1. **Opacity and Deception**: If AI-generated content mimics empathy or authority, should audiences be informed that a machine is behind the message? Without transparency, persuasion can easily edge into deception.
2. **Hyper-Personalization**: AI can detect and exploit micro-vulnerabilities (e.g., insecurity, loneliness). This raises questions similar to those posed by **Zuboff (2019)** in *The Age of Surveillance Capitalism*: where does tailoring end and coercion begin?
3. **Consent Erosion**: Traditional marketing respects opt-ins and clear disclosures. But AI can infer preferences and emotions passively—potentially violating the spirit, if not the letter, of informed consent.
4. **Amplification of Bias**: As noted by **Crawford (2021)** in *Atlas of AI*, data-driven systems often reinforce existing social inequalities. A neuromarketing agent trained in biased inputs may

replicate harmful stereotypes or exclusionary narratives, all in the name of efficiency.

To counter these risks, the chapter proposes an "Ethical Cortex" for neuromarketing AI—seven guiding principles adapted from both neuroscience and AI ethics:

- **Transparency**: Disclose when AI is used and how.
- **Autonomy**: Empower users to opt out or modify engagement.
- **Non-maleficence**: Avoid inducing fear or anxiety without a legitimate user benefit.
- **Beneficence**: Use persuasion to enhance well-being, not just consumption.
- **Justice**: Ensure access and fairness across demographics.
- **Accountability**: Log and review persuasive interventions.
- **Explainability**: Make algorithmic decisions understandable to humans.

Chapter 10 concludes with a sober yet optimistic view: while the tools of influence have evolved, the ethical imperative remains timeless—not to harm. In the hands of conscientious designers, Neuromarketing AI can enhance clarity, trust, and relevance. But without ethical design, it risks becoming a mirror of our worst impulses—optimized at scale.

A Quick Primal Exercise for You

Before you deploy any neuro-optimized message, pause and run this simple test: Would I feel comfortable if my own parent, child, or best friend were on the receiving end of this persuasion? Strip away the analytics. Forget the conversion metrics. Consider the emotional tone of your message. Does it respect the autonomy of the person on the other side? Or does it subtly exploit fear or vulnerability? Your brain knows the answer—trust its signal.

Final Thought: Care, Compassion, and Compass

The power to influence is also the power to respect. In the age of

Neuromarketing AI, the most responsible marketers will not be those who reach the primal brain, they will be those who reach it with care. When technology enables us to communicate directly with the subconscious, the moral question becomes louder, not softer. The future of persuasion depends not just on what we *can* do, but on what we *choose* to do. Ethics, in this light, is not a limit. It is a compass and the most human trait we can encode.

CHAPTER 10: What Comes Next?

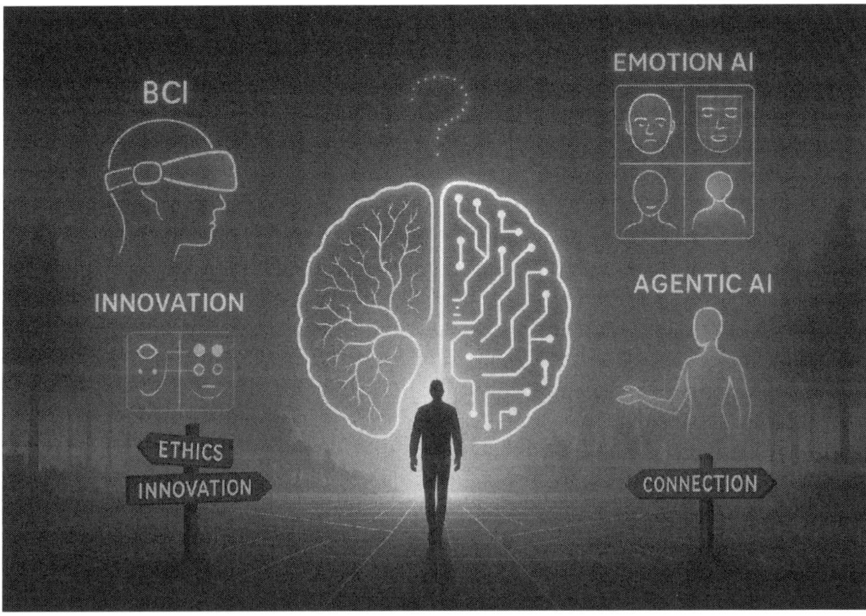

The transition to machine intelligence is a moment of maximum risk and maximum opportunity.

Nick Bostrom, Ph.D.
AI Philosopher

As we stand at the edge of a profound technological and ethical inflection point, the convergence of neuroscience and artificial intelligence invites us to imagine—and build—an entirely new interface between the mind and machine. This last chapter is both a telescope into the future and a map of the decisions we must navigate today. With the emergence of Agentic AI, BCI technology (brain-computer interface), Emotion AI, and intelligent systems that respond to neural signals, the stakes for marketers, scientists, and policymakers have never been higher (Yuste et al., 2017).

The Rise of the Neuromarketing Agents

Artificial Intelligence has evolved from a tool for automation to a partner in cognition. In this concluding chapter, we explore the frontier of this evolution: the emergence of **Agentic AI**—a category of intelligent systems capable of goal-oriented, context-aware, and multi-modal persuasion. More than just software that generates text or answers prompts, Agentic AI behaves like an agent. It sets goals, makes autonomous decisions, and executes tasks aligned with human intent and contextual relevance. In the world of neuromarketing, this is a seismic shift. I also teach Agentic AI at Johns Hopkins and strongly believe in its potential to transform and accelerate the growth of neuromarketing AI radically.

From Generation to Intention: A New Paradigm

To appreciate what Agentic AI brings to the persuasion table, we must first understand its predecessors. Generative AI, like GPT-4 and 5, is designed to predict and produce human-like language (or images, or code) based on vast training datasets. It is reactive by design. It responds to prompts and queries with high fluency and increasing coherence. It is powerful, but it lacks memory, an agenda, and sustained goals.

Then came the Custom GPT. These are tailored instances of generative models, infused with domain-specific knowledge or behavioral preferences. A Custom GPT might be explicitly trained in legal advice, financial planning, or, in our case, neuromarketing strategies. While they are better aligned with context, they remain reactive. They do not initiate action, monitor progress, or adapt strategies over time. This is where Agentic AI steps in—and steps ahead.

Agentic AI is *agenda-driven intelligence*. It is proactive, persistent, and adaptive. It does not just wait for instructions—it acts based on goals. If you tell an Agentic AI to "optimize conversions for Campaign X using neuromarketing principles," it does not just write a headline and stop. It assembles prompts, tests variants, tracks outcomes, and revises strategies across time and media. It operates like a neuromarketing consultant that never sleeps.

Key Capabilities of Agentic AI

1. **Goal Orientation**

Agentic AIs are built around *intent*. They accept objectives, break them into sub-tasks, and autonomously pursue them. These goals are not just linguistic, they are behavioral, measurable, and strategic. This makes Agentic AI ideal for managing entire persuasive journeys, from audience segmentation to neuro-aligned messaging to iterative A/B testing.

2. **Contextual Memory**

Unlike typical generative tools, Agentic AI possesses working memory. It tracks past actions, audience behavior, and environmental feedback. It remembers what worked and what did not making its persuasion strategies not just more personalized, but also increasingly optimized.

3. **Multi-Modal Adaptability**

Agentic AI operates across text, voice, visuals, and data. It can write an emotionally optimized landing page, generate a facially expressive avatar to deliver it, and use biometric feedback to tune its delivery—all without human micromanagement. It is a full-spectrum influence orchestration.

4. **Autonomous Iteration**

The hallmark of intelligence is learning. Agentic AI can simulate, test, and refine its persuasive strategies. It draws on neuromarketing heuristics, NeuroQuadrants™-specific frameworks, and real-time feedback to evolve its messaging. This means the longer it runs, the sharper its impact becomes.

The Neuromarketing Agent in Action

Let us envision a neuromarketing agent powered by Agentic AI. Suppose a wellness brand wants to launch a stress-reduction app to a mid-career, high-burnout demographic. Here is how the Neuromarketing Agent might operate:

- **Step 1: Audience Scanning**
  The agent identifies NeuroQuadrants™ profiles; this audience falls into NeuroQuadrantsTM III (long-term pain avoidance) and NeuroQuadrantsTM IV (long-term pleasure-seeking). They are risk-averse but aspirational.
- **Step 2: Message Crafting**
  Drawing on NeuroPrompting™ teachings, the agent generates content that begins with stress and pain, differentiates through neuroscience-backed relief, and paints a vision of clarity, productivity, and well-being. It generates multiple variants—textual, visual, and even audio.
- **Step 3: Real-Time Testing**
  Using biometric feedback tools (HRV, skin conduction, and EEG data will soon be widely available via wearables), the agent tests which creatives increase attention, emotional engagement, and recall.
- **Step 4: Iteration**
  The agent refines the message hourly, adjusting colors, pacing, language tone, and calls to action—all without waiting for a human prompt.
- **Step 5: Deployment Across Channels**
  It formats the optimized message for email, landing pages, SMS, and voice assistants, modulating each for neural salience based on platform constraints.
  This is persuasion not as a campaign, but as an **ecosystem** that optimizes and aligns with the brain.

Why Neuromarketing Needs Agentic AI

Neuromarketing is complex. It encompasses multiple levels of influence, including sensory, emotional, cognitive, and behavioral. Generative tools can echo into this complexity, but they cannot manage it. Agentic AI is uniquely suited for neuromarketing because:

- It can embed **primal triggers** at scale
- It can track NeuroQuadrants™**-based segmentation** in real time
- It personalizes persuasion down to the neuro-metric level
- It runs 24/7 without fatigue or distraction

- It obeys neurological constraints—like cognitive load and attention span—because it has been trained to recognize them

With Agentic AI, neuromarketing becomes precision influence—intelligent, adaptive, and scalable.

How It Differs from Custom GPT

While both Custom GPTs and Agentic AIs can be built on foundational models, their architectures and behaviors differ profoundly:

Table 1: Custom GPT and Agentic AI

Feature	Custom GPT	Agentic AI
Purpose	Generate context-aligned output	Achieve defined outcomes autonomously
Memory	Session-limited or static	Long-term, dynamic, context-aware
Prompt Dependency	Fully prompt-driven	Goal- and task-driven
Initiative	Never initiates	Initiates based on objective tracking
Modality	Mostly text-based	Multi-modal (text, image, voice, data)
Persuasive Strategy	Requires human orchestration	Orchestrates and adjusts itself
Evolution	Requires retraining or re-prompting	Evolves via feedback and performance data

In simple terms, Custom GPT is a responsive expert. Agentic AI is a persuasive partner.

Ethical Implications for Neuromarketing Agents

As always, with power comes responsibility. Agentic AIs can simulate empathy, urgency, and authority—traits that, if used without ethical guardrails, can slip into manipulation. Chapter 9 already stressed the importance of neuromarketing principles grounded in respect for autonomy, transparency, and non-coercive influence.

Clear limits must also govern an ethically trained Neuromarketing Agent: it should inform and align, rather than deceive or pressure. This includes designing messages with cognitive clarity, signaling promotional intent, and providing opt-out options. Consent must remain sacred—even in the age of hyper-personalized persuasion.

The Future Is Agentic

As we close this book, we stand at the edge of a profound convergence: neuroscience, artificial intelligence, and persuasion science are no longer siloed. They are merging into a new discipline—one where influence is not only predictable, but programmable.

Agentic AI will not replace human marketers. It will amplify their ability to understand, adapt, and emotionally resonate with audiences at scale. For neuromarketers, this is both a challenge and an invitation: to design systems that not only speak to the brain, but also listen to it—patiently, persistently, and ethically.

And so, the journey of persuasion does not end here. It evolves—from neuron to agent, from message to mission.

Brain-Computer Interfaces and Emotion AI

Brain-computer interfaces (BCI) are no longer speculative science fiction. A quarter of a century ago, BCI research was a niche pursuit confined to a handful of specialized labs. Today, it has evolved into a robust and expanding discipline, uniting hundreds of researchers, engineers, and clinicians worldwide (Hassabis et al., 2017). This increasingly collaborative and interdisciplinary community is actively tackling core challenges and unlocking the transformative promise of BCI technologies.

Companies like Neuralink and Kernel have demonstrated prototypes that read electrical activity from the brain and translate it into machine commands. While early applications focus on motor control and therapeutic rehabilitation, the broader implications for human-computer interaction are staggering (Morin, 2023).

Imagine a world where cognitive load is monitored in real-time to reduce digital fatigue, where advertising is tailored not just to demographics but to moment-by-moment neural states, and where communication bypasses language altogether. As BCIs transition from the lab to the market, they will disrupt education, media, healthcare, and, indeed, marketing (Yuste et al., 2017).

BCIs could also unlock new modalities of experience. In future classrooms, for example, a teacher might receive real-time feedback on students' attention levels via neural dashboards. Medical professionals could use BCIs to communicate with patients who are non-verbal. Musicians might compose music by imagining melodies. These possibilities enhance both function and expression (Hassabis et al., 2017).

Athletes may receive cognitive training in real-time, enhancing not only muscle memory but also anticipatory thinking. Air traffic controllers could offload stressful monitoring tasks to AI systems that gauge their cognitive fatigue and recommend micro-rests or automated interventions. In immersive storytelling, audiences could shape narratives via neural feedback, creating hyper-personalized cinema or gaming experiences.

Emotion AI, meanwhile, is already at play. From Amazon's Alexa detecting frustration in a user's voice to customer service platforms that adapt based on facial micro-expressions, these systems are designed to capture affective states and adjust responses accordingly (Zuboff, 2019). In retail, AI can sense whether shoppers are confused or excited. In politics, AI-enhanced sentiment analysis is being used to refine messages in real-time.

Even more nuanced applications are emerging. AI is being trained to recognize complex emotional states—such as ambivalence, subtle distrust, and empathetic resonance—beyond basic facial expressions. This opens up enormous potential for therapeutic use, as AI companions in mental health support could detect emotional instability or distress before humans are even aware of it themselves (Ienca & Andorno, 2017).

In entertainment and education, Emotion AI can modulate experiences based on users' real-time reactions, allowing for adaptive storytelling or customized learning environments. These systems might track emotional arcs across long-term interactions, learning the patterns of individual users and responding with deep personalization (Schüll, 2012).

But with this power comes danger. Who owns the right to your emotions? What if affective data is used not to support, but to steer behavior—toward purchases, beliefs, or decisions that serve the persuader, not the persuaded? Emotion AI and BCIs must be developed with strict ethical boundaries that respect autonomy and identity (Ienca & Andorno, 2017).

We are entering a future where emotional experiences can be encoded and manipulated with unprecedented precision. Without rigorous safeguards, marketers may find themselves tempted to optimize for behavioral outcomes at the expense of human dignity (Zuboff, 2019).

Regulatory Futures and the Ethical AI Marketer

Neuroscience and AI operate in a domain that often precedes regulation. Just as social media outpaced legal oversight for over a decade, neurotechnologies risk galloping ahead of governance unless checked early (Yuste et al., 2017).

International scholars, such as Rafael Yuste, and ethicists, including Ienca and Andorno, have proposed a framework for "neurorights"—encompassing cognitive liberty, mental privacy, psychological continuity, and fair access to mental augmentation. These concepts are gaining traction in global policy circles, from Chile's constitutional reforms to UN advisory boards (Ienca & Andorno, 2017).

Neurorights also highlight critical cultural differences in how the brain is treated. Some societies may prioritize mental privacy over optimization; others may embrace augmentation. This creates a dynamic legal landscape where marketers must adapt strategies across borders while maintaining ethical consistency.

Ethical marketers will need to develop new toolkits, including consent architectures that evolve with the user's state, transparency dashboards that display what data is being used and how, and impact assessments that measure not only profit but also psychological safety.

A comprehensive code of conduct may emerge, not just for tech companies but for anyone engaging in persuasive AI using agents. Much like bioethics in medicine, neuroethics in marketing must become institutionalized embedded into hiring, product design, and even shareholder reports (Ienca & Andorno, 2017).

This new landscape also requires ethical auditing frameworks. Tools must be designed to test not only bias and fairness but also emotional safety. Persuasive content should be evaluated for its psychological impact—are users more anxious after engaging with it?

Do they feel respected? Ethics must become part of the metrics dashboard (Morin, 2023).

The ethical marketer of tomorrow will not just be fluent in neuroscience but literate in philosophy. They will design campaigns that seek resonance, not coercion. They will understand that the brain is a computational organ but a meaning-making machine that deserves dignity (Hassabis et al., 2017).

Future-Proofing Human Connection

As AI becomes increasingly personal and persuasive, we must ask a profoundly human question: Are we building systems that connect—or systems that manipulate? (Zuboff, 2019). The future of communication will not be written in code alone, but in values. To preserve trust, designers of persuasion AI tools must prioritize what we call ethical friction: the small pauses, disclosures, and choices that give users space to reflect. Not every recommendation needs a "yes." Not every decision should be optimized for speed (Morin, 2023).

We should build AI that fosters introspection, emotional resilience, and curiosity. Imagine interfaces that encourage users to explore diverse viewpoints, which highlight when their cognitive load is too high, or that suggest a break instead of another dopamine hit (Schüll, 2012).

Future-proofing also means rethinking what connection means. In digital environments, deep empathy is rare. But it does not have to be. I believe that neuromarketing AI can amplify our capacity for empathy if trained to prioritize human flourishing over engagement metrics. Chatbots that promote conflict resolution, avatars that de-escalate anger, or sentiment-driven content that encourages reflection—these are tools that enhance rather than replace human values (Hassabis et al., 2017).

Meanwhile, smart wearables might help families stay attuned to one another's moods across distances. Teachers could receive emotional trend lines for classrooms, helping them identify and respond to stress in ways that support learning. Community apps might aggregate shared affective data to build collective awareness of neighborhood morale (Morin, 2023).

Neuromarketing AI, if built well, could become a mirror of our best selves—not a tool to shape people into predictable patterns but to support their growth, autonomy, and connection (Yuste et al., 2017).

Chapter Summary

Chapter 10 brings Neuromarketing AI to a forward-looking close, charting the path from today's neuroscience-informed persuasion to tomorrow's autonomous influence systems. Where earlier chapters focused on the convergence of brain science and generative technologies, this concluding chapter zooms out to consider the trajectory of that fusion—where it is going, what it might become, and what it demands of us.

The chapter begins by reminding us that a better understanding has driven every revolution in marketing: first of demographics, then behavior, and now, cognition. With neuromarketing and AI converging, we are entering an era where machines can simulate persuasive reasoning, emotional nuance, and adaptive learning—all guided by brain-based models. But understanding is different from wisdom. The chapter calls on practitioners to be architects of alignment, not just automation.

At the core of this vision is a new paradigm introduced in this edition: **Agentic AI**. Unlike reactive chatbots or static GPT tools, Agentic AI is goal-driven, memory-equipped, and context-aware. It does not just generate content—it pursues outcomes. This makes it the engine behind a new class of intelligent systems: **Neuromarketing Agents**. These agents integrate models like NeuroPrompting™, and real-time behavioral feedback to act as autonomous persuaders—shaping messages, segmenting audiences, testing variants, and learning from neural cues.

Chapter 10 explores how these agents will be embedded into everything from campaign dashboards to virtual assistants, guiding not just what to say, but how and when to say it for maximal resonance. They will identify the optimum NeuroQuadrants™ in real time, adapting to motivational states that shift with context. Persuasion, once static, becomes a dynamic dialogue—calibrated second by second.

This evolution carries profound implications. On the one hand, marketers gain unprecedented precision, scalability, and personalization. On the other hand, the ethical stakes intensify. If machines can anticipate decisions before humans are conscious, the need for guardrails—such as transparency, consent, and emotional integrity—becomes non-negotiable. The book ends with a bold, hopeful note: that Neuromarketing AI is a philosophy—an invitation to speak to the brain with empathy and rigor. It positions practitioners not as manipulators, but as translators between biological truth and digital experience. With the rise of intelligent agents, persuasion itself evolves—not toward coercion, but toward *co-creation.*

The future will not be written by machines alone. It will be shaped by the values we encode, the models we trust, and the signals we choose to follow. The next great frontier of marketing is not just smarter AI. It is more human neuroscience, wielded wisely.

A Quick Primal Exercise for You

Close your eyes and imagine your future self—one year from now—collaborating daily with a Neuromarketing Agent. What tasks would you handle? Which decisions would you still reserve for yourself? What fears do you feel about delegating influence? What excites you about the potential? Jot down your gut reactions. These are not just musings; they are your primal brain signaling its readiness, or resistance, to a new cognitive alliance. Let those instincts shape how you build or adopt the agents of tomorrow.

Final Thought: Respect

As we step into a future shaped by Neuromarketing Agents and Agentic AI, the line between human and machine influence blurs—but never vanishes. The brain remains our compass. It reminds us that persuasion is a relationship: between message and mind, between intent and impact. Whether delivered by a person or a protocol, influence only works when it resonates with what makes us human: emotion, meaning, and trust. The final frontier of persuasion is about building more respectful dialogues. In that future, the brain is a partner.

CONCLUSION

If you have made it this far, I want to say something rare in a field overrun with conversion metrics and clickthrough rates: thank you for your attention.

Not the kind of attention stolen by algorithmic bait, but the rare, focused kind that travels from your brainstem up to your neocortex. That is the journey we have been on together: from the unconscious to the conscious, from the subterranean machinery of the mind to the surface of marketing tactics. Moreover, as someone who has spent most of his life obsessed with decoding why humans say "yes," I can tell you—we have only just begun to understand the terrain.

As you are now aware, my fascination with persuasion did not begin in academia. It started much earlier, in the back of one of the mobile shops my father built in western France in the sixties. Imagine a traveling storefront—a fusion of utility and theater—where the scent of cheese mingled with diesel fuel and negotiation happened between the blink of an eye and the gesture of a hand. I did not know it at the time, but I was already studying neuromarketing. Those trucks were experiments. My father did not know neuroscience, but he knew how to grab attention, build trust, and close the sale—long before anyone said a word.

Later, I pursued neuroscience, a gift my father could not afford. Degrees, labs, EEG machines, thousands of biometric recordings, hundreds of brains under the lens. And what did I find? That we are still, at our core, primal creatures navigating a modern maze. Our behavior is not controlled by what we think, but by what we *feel before we think.* And yet, most marketing still insists on shouting logic into the wind, hoping it lands somewhere useful.

That is why this book exists.

What Neuromarketing AI Can Do—and What It *Should* Do

We stand at the crossroads of profound opportunity and enormous risk. Artificial intelligence can now write, speak, translate, imitate, and even charm. But can it persuade with integrity? Can it stir the heart without hacking it?

Let us be brutally honest. AI, as it stands today, is a toddler with superpowers. It can process petabytes of data, but does not yet grasp what it means to be human. It can imitate emotional expression, but it does not convey genuine feelings. That is where we come in—not to control AI, but to teach it *where the soul lives* in a message.

Neuromarketing AI, as I define it, is not just the fusion of neuroscience and artificial intelligence. It serves as the ethical compass guiding how machines should interact with minds. It is an insistence that technology *earn* trust, not just harvest it. Moreover, it is a framework for designing communication that honors both our evolutionary wiring and our aspirational intelligence.

When I teach this discipline, I often show a simple slide: "Speak to the primal brain first, or do not bother." It sounds extreme. But think about it. Have you ever changed your mind because someone gave you more bullet points? Probably not. But you have changed your behavior because something felt right, because it mattered to you personally, or because it hit you in a moment of emotional clarity. That is not manipulation. That is alignment with the way the brain works.

Ethical Influence is Not an Oxymoron

Let us pause here. In this age of digital noise, AI hype, and data-driven everything, it is tempting to treat the brain like a computer: input, process, output. But our brains are not spreadsheets. They are stories in motion. They remember based on emotion. They decide based on fear and hope. And they connect not to code, but to care.

As we develop persuasive machines, we must ask: Who are we serving? Are we helping people make better decisions—or just faster ones? Are we revealing truth—or simulating it so well that we no longer know the difference?

In our world of predictive models and personalized prompts, the line between persuasion and manipulation is razor-thin. That is why ethics is not just a postscript to this book—it is the spine.

In my decades of research and consulting, I have always insisted on three non-negotiables:

1. **Informed transparency**: If we are measuring the brain, people should know what we are doing, why, and how it benefits them.
2. **Mutual benefit**: Ethical persuasion is a win-win. If only one party wins, it is no longer persuasion. It is coercion.

3. **Guardrails, not gimmicks**: Persuasion should never exploit vulnerabilities—whether through AI-enhanced fear mongering or emotionally engineered clickbait.

AI is powerful. But power without empathy is a weapon. With empathy, it becomes a tool for elevation.

A New Generation of Persuaders

If you are a marketer, designer, writer, founder, or data scientist, this is your moment. You are not just creating campaigns. You are shaping how the world thinks. And with the tools you have explored in this book—primal triggers, neural models, biometric feedback, and NeuroMap®—you are now equipped to do so with scientific precision and human grace.

Use these tools not to shout louder, but to listen deeper. Not to manipulate, but to resonate. Not to reduce humans to segments, but to honor the astonishing complexity of how we decide.

I still remember the first time a client saw their EEG results from a campaign. The excitement was instant. "So, this is what the brain really thinks?" they asked. Yes. And it was nothing like what the focus group said. That moment never gets old. It reminds me that beneath the personas, the KPIs, and the dashboards, there is a living, feeling brain—trying to make sense of a noisy world. Our job is to make that world more transparent, more humane.

Let Us Leave the World with Better Questions

If there is one gift I hope this book has given you, it is not just answers. It is a better question.

- What if we designed AI that respected the brain's limitations, instead of exploiting them?
- What if we measured not just attention, but intention?
- What if persuasion could feel like empowerment?

I do not have all the answers. But I do have a compass. And I have tried, in these pages, to convey it to you. You now know how the brain decides. You know how the primal brain craves clarity, emotion, relevance, and simplicity. You know that the best AI is not the most human-sounding—it is the most human-centered.

So, use what you have learned here. Use it to design, to test, to build, to inspire. But use it to connect—with integrity, with empathy, and with the brain in mind. Because when we persuade ethically, we do not just sell better. We serve better.

Here is the final coda—designed to echo the book's most humanistic message while leaving the reader with a clear sense of responsibility and inspiration. It can serve as both a poetic ending and a rallying cry.

Coda: The New Persuaders' Manifesto

Let me end not with a summary, but with a challenge. We stand on the edge of something immense, a revolution not just in marketing or artificial intelligence, but in meaning. We now have the tools to speak directly to the decision-making core of the human brain. The question is—what will we say?

Will we flood that core with noise, fear, and a sense of urgency? Or will we craft messages that elevate, clarify, and connect?

The answer is not in the algorithm. It is in us. Persuasion, in its highest form, is ultimately not about control. It is about alignment. Between message and mind. Between action and ethics. Between machine logic and human magic.

So be bold. Build GPTs that understand not only syntax but also narrative. Design campaigns that feel like conversations, not commands. Run experiments not just to optimize, but to empathize.

And remember this:

- The primal brain does not want perfection.
- It wants the truth it can feel.
- It wants clarity it can trust.
- It wants messages that earn their place in memory.

This book has given you the science behind persuasion. It has given you the map. \Now it is your turn to chart the terrain. To step into the role of the new persuader—not a manipulator, not a magician, but a guide.

A guide who understands that while machines may learn faster, only humans can teach what matters most: meaning, empathy, and ethical intention.

The brain has always been the final frontier.

Now, let us explore it with wisdom.

And just maybe, if we do this right, we will create a world where persuasion feels less like pressure—and more like possibility.

Dr. Christophe Morin, September 21, 2025.
Kona, Hawaii

REFERENCES

Agee, T., & Martin, B. A. S. (2001). Planned or impulse purchases? How to create effective infomercials. Journal of Advertising Research, 41(6), 35–42.

Al-Nafjan, A., Hosny, M., Al-Ohali, Y., & Al-Wabil, A. (2022). EEG-based emotion recognition: A review. Sensors, 22(7), 2551. https://doi.org/10.3390/s22072551

Ariely, D. (2008). Predictably irrational: The hidden forces that shape our decisions. HarperCollins.

Ariely, D., & Berns, G. S. (2010). Neuromarketing: The hope and hype of neuroimaging in business. Nature Reviews Neuroscience, 11(4), 284–292. https://doi.org/10.1038/nrn2795

Aristotle. (ca. 350 B.C.E./1991). Rhetoric (G. A. Kennedy, Trans.). Oxford University Press. (Original work published ca. 350 B.C.E.)

Asch, S. E. (1951). Effects of group pressure upon the modification and distortion of judgments. In H. Guetzkow (Ed.), Groups, leadership and men (pp. 177–190). Carnegie Press.

Barrett, L. F. (2017). How emotions are made: The secret life of the brain. Houghton Mifflin Harcourt.

Başar, E., Başar-Eroglu, C., Karakaş, S., & Schürmann, M. (2001). Gamma, alpha, delta, and theta oscillations govern cognitive processes. International Journal of Psychophysiology, 39(2–3), 241–248. https://doi.org/10.1016/S0167-8760(00)00145-8

Baumeister, R. F., & Tierney, J. (2011). Willpower: Rediscovering the greatest human strength. Penguin.

Bernays, E. L. (1928). Propaganda. Liveright.

Berns, G. S. (2005). Satisfaction: The science of finding true fulfillment. Henry Holt.

Cambria, E., & White, B. (2014). Jumping NLP curves: A review of natural language processing research. IEEE Computational Intelligence Magazine, 9(2), 48–57. https://doi.org/10.1109/MCI.2014.2307227

Chaiken, S. (1980). Heuristic versus systematic information processing and the use of source versus message cues in persuasion. Journal of Personality and Social Psychology, 39(5), 752–766. https://doi.org/10.1037/0022-3514.39.5.752

Cogito. (2020). Empathy at scale: Enhancing customer experience with AI. Cogito Corporation. [White Paper].

ContextSDK. (2024, December 23). Understanding the evolution of direct response advertising. ContextSDK. [White Paper].

Cozolino, L. (2010). The neuroscience of psychotherapy: Healing the social brain (2nd ed.). W. W. Norton & Company.

Cowen, A. S., Elfenbein, H. A., Laukka, P., & Keltner, D. (2019). Mapping 24 emotions conveyed by brief human vocalization. American Psychologist, 74(6), 698–712. https://doi.org/10.1037/amp0000399

Cowie, R., Douglas-Cowie, E., Tsapatsoulis, N., Votsis, G., Kollias, S., Fellenz, W., & Taylor, J. G. (2001). Emotion recognition in human-computer interaction. IEEE Signal Processing Magazine, 18(1), 32–80. https://doi.org/10.1109/79.911197

Critchley, H. D. (2002). Electrodermal responses: What happens in the brain. The Neuroscientist, 8(2), 132–142. https://doi.org/10.1177/107385840200800209

Damasio, A. R. (1994). Descartes' error: Emotion, reason, and the human brain. Putnam.

Davidson, R. J. (1992). Anterior cerebral asymmetry and the nature of emotion. Brain and Cognition, 20(1), 125–151. https://doi.org/10.1016/0278-2626(92)90065-T

Dodge, R., & Cline, T. (1901). The angle velocity of eye movements. Psychological Review, 8(2), 145–157. https://doi.org/10.1037/h0076100

Dunbar, R. (1998). The social brain hypothesis. Evolutionary Anthropology, 6(5), 178–190. https://doi.org/10.1002/(SICI)1520-6505(1998)6:5
<178::AID-EVAN5>3.0.CO;2-8

Eagleman, D. (2011). Incognito: The secret lives of the brain. Pantheon.

Ekman, P., & Friesen, W. V. (1978). Facial action coding system: A technique for the measurement of facial movement. Consulting Psychologists Press.

Escalas, J. E. (2004). Narrative processing: Building consumer connections to brands. Journal of Consumer Psychology, 14(1–2), 168–180. https://doi.org/10.1207/s15327663jcp1401&2_19

European Commission. (2020). Ethics guidelines for trustworthy AI. High-Level Expert Group on Artificial Intelligence. Publications

Office of the European Union. https://ec.europa.eu/futurium/en/ai-alliance-consultation/guidelines

Festinger, L. (1957). A theory of cognitive dissonance. Stanford University Press.

Floridi, L., & Cowls, J. (2019). A unified framework of five principles for AI in society. Harvard Data Science Review, 1(1). https://doi.org/10.1162/99608f92.8cd550d1

Gazzaniga, M. S. (2018). The consciousness instinct: Unraveling the mystery of how the brain makes the mind. Farrar, Straus and Giroux.

Geisler, F. C. M., Vennewald, N., Kubiak, T., & Weber, H. (2013). Heart rate variability and emotional regulation. Biological Psychology, 93(1), 379–386. https://doi.org/10.1016/j.biopsycho.2013.04.005

Gladwell, M. (2005). Blink: The power of thinking without thinking. Little, Brown.

Glanzer, M., & Cunitz, A. R. (1966). Two storage mechanisms in free recall. Journal of Verbal Learning and Verbal Behavior, 5(4), 351–360. https://doi.org/10.1016/S0022-5371(66)80044-0

Green, M. C., & Brock, T. C. (2000). The role of transportation in the persuasiveness of public narratives. Journal of Personality and Social Psychology, 79(5), 701–721. https://doi.org/10.1037/0022-3514.79.5.701

Greenwald, A. G., McGhee, D. E., & Schwartz, J. L. K. (1998). Measuring individual differences in implicit cognition: The implicit association test. Journal of Personality and Social Psychology, 74(6), 1464–1480. https://doi.org/10.1037/0022-3514.74.6.1464

Harmon-Jones, E., & Gable, P. A. (2018). On the role of asymmetric frontal cortical activity in approach and withdrawal motivation: An updated review. Biological Psychology, 140, 35–49. https://doi.org/10.1016/j.biopsycho.2018.01.008

Hassabis, D., Kumaran, D., Summerfield, C., & Botvinick, M. (2017). Neuroscience-inspired artificial intelligence. Neuron, 95(2), 245–258. https://doi.org/10.1016/j.neuron.2017.06.011

Hasson, U., Ghazanfar, A. A., Galantucci, B., Garrod, S., & Keysers, C. (2012). Brain-to-brain coupling: A mechanism for creating and sharing a social world. Trends in Cognitive Sciences, 16(2), 114–121. https://doi.org/10.1016/j.tics.2011.12.007

Hovland, C. I., Janis, I. L., & Kelley, H. H. (1953). Communication and persuasion: Psychological studies of opinion change. Yale University Press.

Ienca, M., & Andorno, R. (2017). Towards new human rights in the age of neuroscience and neurotechnology. Life Sciences, Society and Policy, 13(5). https://doi.org/10.1186/s40504-017-0050-1

Jones, T., & Laker, L. (2023). Neural decoding of emotional impact in advertising. Neuroscience & Marketing Insights Journal, 18(3), 215–229.

Jordan, M. I., & Mitchell, T. M. (2015). Machine learning: Trends, perspectives, and prospects. Science, 349(6245), 255–260. https://doi.org/10.1126/science.aaa8415

Kahneman, D. (2011). Thinking, fast and slow. Farrar, Straus and Giroux.

Kahneman, D., & Tversky, A. (1979). Prospect theory: An analysis of decision under risk. Econometrica, 47(2), 263–291. https://doi.org/10.2307/1914185

Kapoor, A., Burleson, W., & Picard, R. W. (2018). Automatic prediction of frustration. International Journal of Human-Computer Studies, 65(8), 724–736.

Kellogg Insight. (2021, March 1). TV advertising is usually not worth it. Northwestern University. https://insight.kellogg.northwestern.edu

Kernel. (2021). The future of brain-computer interfaces. Kernel. [Press Release].

Klimesch, W. (1999). EEG alpha and theta oscillations reflect cognitive and memory performance: A review and analysis. Brain Research Reviews, 29(2–3), 169–195. https://doi.org/10.1016/S0165-0173(98)00056-3

Klucharev, V., Hytönen, K., Rijpkema, M., Smidts, A., & Fernández, G. (2009). Reinforcement learning signal predicts social conformity. Neuron, 61(1), 140–151. https://doi.org/10.1016/j.neuron.2008.11.027

LeCun, Y., Bengio, Y., & Hinton, G. (2015). Deep learning. Nature, 521(7553), 436–444. https://doi.org/10.1038/nature14539

LeDoux, J. (1996). The emotional brain. Simon & Schuster.

Lewinski, P., den Uyl, T. M., & Butler, C. (2014). Automated facial coding: Validation of basic emotions and FACS AUs in FaceReader. Journal of Neuroscience, Psychology, and Economics, 7(4), 227–236. https://doi.org/10.1037/npe0000028

Lieberman, M. D. (2013). Social: Why our brains are wired to connect. Crown.

Lisman, J. E., & Grace, A. A. (2005). The hippocampal-VTA loop: Controlling the entry of information into long-term memory. Neuron, 46(5), 703–713. https://doi.org/10.1016/j.neuron.2005.05.002

Luck, S. J. (2014). An introduction to the event-related potential technique (2nd ed.). MIT Press.

MacLean, P. D. (1990). The triune brain in evolution: Role in paleocerebral functions. Springer.

Mar, R. A., & Oatley, K. (2008). The function of fiction is the abstraction and simulation of social experience. Perspectives on Psychological Science, 3(3), 173–192. https://doi.org/10.1111/j.1745-6924.2008.00073.x

Martin, B. A. S., Bhimy, A., & Agee, T. (2002). Infomercials and advertising effectiveness: An empirical study. Journal of Consumer Marketing, 19(6), 468–480. https://doi.org/10.1108/07363760210444872

McClure, S. M., Li, J., Tomlin, D., Cypert, K. S., Montague, L. M., & Montague, P. R. (2004). Neural correlates of behavioral preference for culturally familiar drinks. Neuron, 44(2), 379–387. https://doi.org/10.1016/j.neuron.2004.09.019

McDuff, D., Czerwinski, M., & Horvitz, E. (2019). Exploring the role of emotion in human-computer interaction. Proceedings of the 2019 CHI Conference on Human Factors in Computing Systems. https://doi.org/10.1145/3290605.3300341

McGuire, W. J. (1961). The effectiveness of supportive and refutational defenses in immunizing and restoring beliefs against persuasion. Sociometry, 24(2), 184–197. https://doi.org/10.2307/2786067

Morin, C., & Renvoisé, P. (2007). Neuromarketing: Understanding the buy buttons in your customer's brain. Thomas Nelson.

Morin, C., & Renvoisé, P. (2018). The persuasion code: How neuromarketing can help you persuade anyone, anywhere, anytime. Wiley.

Morin, C. (2020). The ultimate neuromarketing research guide. SalesBrain Publishing.

Morin, C. (2021). The serenity code: Neuroscience-based strategies for stress resilience. SalesBrain Publishing.

Morin, C. (2025). Open: A neurospiritual exploration of the self-healing power of your brain. SalesBrain Publishing.

Morin, C. (2025). Open up: The neurospiritual guide to your Enneagram type. SalesBrain Publishing.

Mozziconacci, S. J. (2002). Speech variability and emotion: Production and perception. Speech Communication, 40(1–2), 1–7. https://doi.org/10.1016/S0167-6393(02)00057-5

Murdock, B. B. (1962). The serial position effect of free recall. Journal of Experimental Psychology, 64(5), 482–488. https://doi.org/10.1037/h0045106

Murre, J. M. J., & Dros, J. (2015). Replication and analysis of Ebbinghaus' forgetting curve. PLOS ONE, 10(7), e0120644. https://doi.org/10.1371/journal.pone.0120644

Neurable. (2020). Company research and product announcements. Neurable.

Neurable. (2022). Neurable unveils next-generation brain-computer interface. Neurable. [Company Report].

Niedermeyer, E., & Lopes da Silva, F. (2004). Electroencephalography: Basic principles, clinical applications, and related fields. Lippincott Williams & Wilkins.

Northoff, G., Heinzel, A., de Greck, M., Bermpohl, F., Dobrowolny, H., & Panksepp, J. (2006). Self-referential processing in our brain—a meta-analysis of imaging studies on the self. NeuroImage, 31(1), 440–457. https://doi.org/10.1016/j.neuroimage.2005.12.002

O'Connell, R. G., Dockree, P. M., & Kelly, S. P. (2012). A supramodal accumulation-to-bound signal that determines perceptual decisions in humans. Nature Neuroscience, 15(12), 1729–1735. https://doi.org/10.1038/nn.3248

O'Neil, C. (2016). Weapons of math destruction. Crown Publishing.

OpenAI. (2024). Reinforcement learning from human and AI feedback. OpenAI Technical Report.

Petty, R. E., & Cacioppo, J. T. (1986). The elaboration likelihood model of persuasion. In L. Berkowitz (Ed.), Advances in experimental social psychology (Vol. 19, pp. 123–205). Academic Press. https://doi.org/10.1016/S0065-2601(08)60214-2

Pfurtscheller, G., & Lopes da Silva, F. H. (1999). Event-related EEG/MEG synchronization and desynchronization: Basic principles. Clinical Neurophysiology, 110(11), 1842–1857. https://doi.org/10.1016/S1388-2457(99)00141-8

Phelps, E. A. (2004). Human emotion and memory: Interactions of the amygdala and hippocampal complex. Current Opinion in Neurobiology, 14(2), 198–202. https://doi.org/10.1016/j.conb.2004.03.015

Picard, R. W. (1997). Affective computing. MIT Press.
Pieters, R., & Wedel, M. (2004). Attention capture and transfer in advertising: Brand, pictorial, and text-size effects. Journal of Marketing, 68(2), 36–50. https://doi.org/10.1509/jmkg.68.2.36.277
Ploog, D. (2003). The place of the triune brain in psychiatry. European Archives of Psychiatry and Clinical Neuroscience, 253(Suppl. 2), II-50–II-55. https://doi.org/10.1007/s00406-007-2002-7
Polich, J. (2007). Updating P300: An integrative theory of P3a and P3b. Clinical Neurophysiology, 118(10), 2128–2148. https://doi.org/10.1016/j.clinph.2007.04.019
Porges, S. W. (2011). The polyvagal theory: Neurophysiological foundations of emotions, attachment, communication, and self-regulation. W. W. Norton & Company.
Potter, M. C., Wyble, B., Hagmann, C. E., & McCourt, E. S. (2014). Detecting meaning in RSVP at 13 ms per picture. Attention, Perception, & Psychophysics, 76(2), 270–279. https://doi.org/10.3758/s13414-013-0605-z
Raichle, M. E., & Gusnard, D. A. (2002). Appraising the brain's energy budget. Proceedings of the National Academy of Sciences, 99(16), 10237–10239. https://doi.org/10.1073/pnas.172399499
Ramsøy, T. Z. (2015). *Introduction to neuromarketing & consumer neuroscience.* Neurons Inc.
Ray, W. J., & Cole, H. W. (1985). EEG alpha activity reflects attentional demands, and beta activity reflects emotional and cognitive processes. Science, 228(4700), 750–752. https://doi.org/10.1126/science.3992243
Realeyes. (2021). The impact of emotional engagement on ad performance. Realeyes. [Research Brief].
Reber, R., Schwarz, N., & Winkielman, P. (2004). Processing fluency and aesthetic pleasure: Is beauty in the perceiver's processing experience? Personality and Social Psychology Review, 8(4), 364–382. https://doi.org/10.1207/s15327957pspr0804_3
Riesenhuber, M., & Poggio, T. (1999). Hierarchical models of object recognition in cortex. Nature Neuroscience, 2(11), 1019–1025. https://doi.org/10.1038/14819
Rizzolatti, G., Fadiga, L., Gallese, V., & Fogassi, L. (1996). Premotor cortex and the recognition of motor actions. Cognitive Brain Research, 3(2), 131–141. https://doi.org/10.1016/0926-6410(95)00038-0

SalesBrain. (2018). Secrets from the persuasion code. SalesBrain Inc.
SalesBrain. (2022). Neuromarketing campaign case studies. SalesBrain Inc.
Sara, S. J. (2009). The locus coeruleus and noradrenergic modulation of cognition. Nature Reviews Neuroscience, 10(3), 211–223. https://doi.org/10.1038/nrn2573
Schüll, N. D. (2012). Addiction by design: Machine gambling in Las Vegas. Princeton University Press.
Schwartz, B. (2004). The paradox of choice: Why more is less. Harper Perennial.
Simonzad, D. (2025, April 15). What is DRTV? Exploring its impact on modern advertising. GoDRTV. https://godrtv.com
Smidts, A., Hsu, M., Sanfey, A. G., Boksem, M. A. S., Ebstein, R. B., Huettel, S. A., ... & Yoon, C. (2014). Advancing consumer neuroscience. Marketing Letters, 25(3), 257–267. https://doi.org/10.1007/s11002-014-9306-1
Smith, M. L., Davidson, R. J., et al. (2014). Frontal asymmetry and consumer response. Journal of Consumer Psychology, 24(1), 132–140. https://doi.org/10.1016/j.jcps.2013.09.002
Somerville, L. H., Heatherton, T. F., & Kelley, W. M. (2006). Anterior cingulate cortex responds differentially to expectancy violation and social rejection. Nature Neuroscience, 9(8), 1007–1008. https://doi.org/10.1038/nn1728
Teixeira, T., Wedel, M., & Pieters, R. (2012). Emotion-induced engagement in internet video advertisements. Journal of Marketing Research, 49(2), 144–159. https://doi.org/10.1509/jmr.10.0209
Thaler, R. H., & Sunstein, C. R. (2008). Nudge: Improving decisions about health, wealth, and happiness. Yale University Press.
Tversky, A., & Kahneman, D. (1981). The framing of decisions and the psychology of choice. Science, 211(4481), 453–458. https://doi.org/10.1126/science.7455683
Venkatraman, V., Dimoka, A., Pavlou, P. A., Vo, K., Hampton, W., Bollinger, B., ... & Winer, R. S. (2015). Predicting advertising success beyond traditional measures: New insights from neurophysiological methods and market response modeling. Journal of Marketing Research, 52(4), 436–452. https://doi.org/10.1509/jmr.13.0593
Venkatraman, V., Dimoka, A., Pavlou, P. A., Vo, K., Hampton, W., Bollinger, B., ... & Winer, R. S. (2024). Predicting advertising

success beyond traditional measures: New insights from neurophysiological methods. Journal of Advertising Research, 64(1), 44–59.

Yarbus, A. L. (1967). Eye movements and vision. Plenum Press.

Yuste, R., & Goering, S. (2017). The ethical challenges of neurotechnology. Nature Biotechnology, 35(12), 1059–1061. https://doi.org/10.1038/nbt.4031

Yuste, R., Goering, S., Arcas, B. A. Y., Bi, G., Carmena, J. M., Carter, A., ... & Wolpaw, J. (2017). Four ethical priorities for neurotechnologies and AI. Nature, 551(7679), 159–163. https://doi.org/10.1038/551159a

Zaki, J., Schirmer, J., & Mitchell, J. P. (2011). Social influence modulates the neural computation of value. Psychological Science, 22(7), 894–900. https://doi.org/10.1177/0956797611411057

Zaltman, G. (2003). How customers think: Essential insights into the mind of the market. Harvard Business School Press.

Zuboff, S. (2019). The age of surveillance capitalism. PublicAffairs.

TABLE OF IMAGES

AUTHOR BIOGRAPHY

Dr. Morin's work is grounded in the belief that most decisions occur below our level of consciousness. He is a globally recognized expert in neuromarketing and a devoted researcher of consciousness. With over two decades at the intersection of neuroscience, psychology, and marketing, Dr. Morin has dedicated his career to decoding how the brain processes messages and makes decisions.

As the co-founder of SalesBrain, the first neuromarketing training agency established in 2002, he co-created and pioneered the NeuroMap® persuasion model, a scientifically grounded framework taught to over 250,000 executives in more than 30 countries.

Using biometric tools such as EEG, GSR, eye tracking, and facial coding, Dr. Morin has confirmed that emotion, rather than logic, is the actual engine of persuasion. His insights continue to reshape how brands craft messages that resonate deeply and ethically.

His latest book, Neuromarketing AI: The New Science of Persuasion in the Age of Intelligent Machines, explores how artificial intelligence can scale the impact of primal brain triggers through ethically engineered content. It introduces the concept of NeuroPrompting and offers a framework for building GPTs that align with human cognitive architecture.

Dr. Morin also teaches at Fielding Graduate University and Johns Hopkins University, where he created multiple AI courses, including the world's only graduate-level course dedicated explicitly to Neuromarketing AI.

Holding a PhD in Media Psychology with a focus on neuropsychology and an MBA, Dr. Morin's research spans decision science, affective neuroscience, consciousness, wellness, and the ethical dimensions of human-machine communication. Through his teaching, research, and writing, he continues to lead the global movement toward a more brain-savvy and ethically conscious approach to influence.

Made in the USA
Columbia, SC
11 October 2025

0bc5abfb-9260-4eac-8d71-ab12225cdde1R01